Praise for *Octopus!*

"[Courage] gamely explores the bustling realm of the octopus . . . entertaining and eye-opening . . . readers will look forward to more of Courage's jauntily elucidating dispatches."

—*Booklist*

"The octopus is like an alternate experiment in intelligent life—sophisticated, alluring, and wholly alien. In her fresh, deeply reported book, Katherine Harmon Courage makes this creature a little less mysterious, but no less marvelous."

—Carl Zimmer, author of *A Planet of Viruses* and *The Tangled Bank: An Introduction to Evolution*

"What is it like to be an octopus? We need to imagine that our arms, all eight of them, can think and that our skin can see. Despite our obvious trouble understanding this invertebrate, no one doubts its intellect. A delightful book about a fascinating creature!"

—Frans de Waal, author of *The Bonobo and the Atheist*

"A well-written, accessible book."

—*Library Journal* (Judith Barnett recommended review)

CURRENT

OCTOPUS!

Katherine Harmon Courage is an award-winning freelance journalist and a contributing editor for *Scientific American*. She has also written for the *New York Times, Wired, Popular Science, Prevention, Gourmet, Nature,* and other publications, and her work was featured in *The Best American Science and Nature Writing 2013*. She lives, works, runs, and gardens in Colorado with her husband. Read more on her Web site www.KatherineCourage.com, and follow her on Twitter at @KHCourage.

OCTOPUS!

The Most Mysterious Creature in the Sea

KATHERINE HARMON COURAGE

CURRENT

CURRENT
Published by the Penguin Group
Penguin Group (USA) LLC
375 Hudson Street
New York, New York 10014

USA | Canada | UK | Ireland | Australia | New Zealand | India | South Africa | China
penguin.com
A Penguin Random House Company

First published in the United States of America by Current,
a member of Penguin Group (USA) LLC, 2013
This paperback edition published 2014

ISBN 978-1-59184-572-0 (hc.)
ISBN 978-1-61723-014-1 (pbk.)

Printed in the United States of America
10 9 8 7 6 5 4 3 2 1

Set in Palatino
Designed by Carla Bolte

To my grandfather

TED ROGERS

the most amazing (human) specimen I've met

Does the complete understanding of a natural phenomenon strip away its miraculous qualities? It is certainly a risk. But it should at least maintain all of its poetry, for poetry subverts reason and is never dulled by repetition. Besides, a few gaps in our knowledge will always allow for a joyous confusion of the mysterious, the unknown, and the miraculous.

—Jean Painlevé, filmmaker, "Mysteries and Miracles of Nature," *Vu*, 1931

CONTENTS

OCTOPUS!

Andreino the octopus muse from the
Laschi biorobotics lab in Livorno, Italy.
(Katherine Harmon Courage)

Introduction

The octopus is a tough beast to grasp. With eight arms, three hearts, camouflaging skin, and a disarmingly sentient look in its eyes, how could it appear anything but utterly alien? The octopus has been beguiling humans for as long as we have been catching it: millennia. Cultures have created octopus-centric creation myths, art, and, of course, cuisine. For all of our ancient fascination and the millions of dollars funneled into modern research, we still have not been able to get a handle on these slippery creatures.

But let's not let that stop *us*. From a fishing boat on the high Spanish seas to a robotics lab in Italy to an octopus distributor in Brooklyn, we will discover what makes the octopus so fascinating—and what it can teach us. (And yes, for good measure, we'll also taste a few of them.) So sit back and relax—fix yourself an octopus's garden cocktail* if you like—as we plunge into the realm of the strong, slick, smart, and occasionally delicious octopus.

The ancient Greeks christened the animal *októpous*, which means, unimaginatively—and slightly inaccurately—"eight foot" (the correct label for the appendages is actually "arms"). And while we're clearing the nomenclatorial air, thanks to this Greek origin, the preferred common plural is *octopuses*—not *octopi*, *octopodes*, or *octopussies*.

Whatever one chooses to call them, there is no doubt that these animals have a strange hold on our imaginations. The stories of monstrous, multiarmed mythical krakens surface periodically throughout

* Octopus's garden cocktail: Shake three parts gin and one part dry vermouth with ice; strain and serve garnished with a baby octopus and a black olive.

literature, legend, art, and film. According to Hawaiian mythology, the octopus is the only living holdover from the world's previous incarnations. Per local Gilbert Islands legend, the octopus god Na Kika is responsible for having pushed the islands up from the bottom of the sea. The strange octopus-dragon-human beast Cthulhu, created by writer H. P. Lovecraft, has appeared in popular culture for nearly a century and was even featured on a few episodes of the television show *South Park*. Pliny the Elder's *Naturalis Historia*, from the first century, describes a giant octopus that weighed in excess of six hundred pounds and took fish from villagers. And a picturesque Italian town called Tellaro, on the coast in Liguria, has erected homages to the giant octopus that reputedly saved the village by ringing the church tower bell to warn of an impending invasion.

In the mid-twentieth century, sporting divers vied to wrest some of the biggest octopuses from the ocean in organized octopus wrestling competitions. More recently, contestants on a Japanese game show tackled a tremendous (if sluggish) specimen in a large tank. The octopus has served as inspiration for at least a couple of comic book super-

Octopus door adornment in Tellaro, Italy.
(Katherine Harmon Courage)

villains who use their intellect and sometimes their additional arms to further their nefarious plots. Octopuses have crept into real-life research labs, helping scientists to better understand the nervous system, build camouflaging materials, and construct soft-bodied robots.

Today, anyone with an Internet connection can view a number of public aquarium octopus webcams and see what the animals are up to *right at this very moment*. Octopus lovers can also join the online *Cephalopod Tea Party* or the discussion at the *Everything Octopus* blog. True devotees log on to the TONMO (*The Octopus News Magazine Online*) community and attend the related semiannual TONMOCON (the TONMO Convention). The octopus has even been the muse for an entire short story anthology (*Suction Cup Dreams*, which includes titles such as "Vulgaris," "A Stranger Returns from an Unexpected Trip to the South China Sea," and "Three-Hearted"). Its likeness crops up in tattoos, as upholstery patterns, on neckties, and even in erotic art—a testament to its odd and far-reaching allure. The singer Fiona Apple demonstrated, in a 2012 music video, that an octopus can even be fashionable head wear. Internet videos showing them strangling sharks or lurching about on land (both feats they are capable of) have made these animals offbeat stars.

Humans have been catching octopuses for thousands of years. We currently drag in more than fifty thousand tons of these muscly animals annually. People have caught them with lures, spears, pot traps, nets, and their bare hands. Culinary strategies are even more varied. You can order them boiled as a Spanish appetizer, baked into a Maltese pizza, pressed and sliced into Italian octopus *soppressata*, grilled as a Greek entrée, raw as Japanese sushi, or even half alive in Korean cuisine. You can also pick them up brined, dried, cured, salted, or frozen. In the narrow streets of some Greek islands, you can literally run right into fresh, locally caught octopuses hanging out on lines to dry in the sun. Although the arms are usually the main dining attraction, the ink, which the animals employ for defense, is still used today for sauces and food coloring.

But many researchers profess to love the octopus not for its meat—er, arms—but for its brains. "They are interesting because they are the smartest invertebrate," says biologist Roland Anderson, who worked with octopuses for more than forty years at the Seattle Aquarium. Okay, so they're smarter than your average garden worm, big deal. But when you consider that more than 95 percent of the creatures on earth are invertebrates, they're facing an awful lot of competition, which comes from a lot of different evolutionary avenues. And octopuses are, after all, mollusks, as are oysters, which, Anderson points out, "don't do much" and aren't terribly intelligent. "Octopuses do so much more," he says. "They are a predator, they go out to find food, they build dens and they modify them, they use tools, they use spatial navigation, they have play behavior, they recognize individual people. So . . . octopuses are pretty smart." And, yes, some of them even make "gardens."

Yet octopuses haven't created extensively manicured estates, built any underwater Atlantises, or developed complex social groups à la humpback whales. But their lack of civilization doesn't mean they aren't smart. As we shall soon find out, they are incredibly intelligent. They can solve simple mazes, open childproof bottles, and even, some argue, use tools. Not too shabby when your closest kin are calamari. Some devotees of Paul, the 2010 FIFA World Cup match–predicting octopus, will contend that they can even be prescient. So popular was the now-deceased Paul, he has inspired numerous imitators, including other octopuses, a llama, a pig, a ferret, an elephant, and various other creatures. He is also the inspiration for a German fortune-telling book, *Das Okrakel: Krake Paul prophezeit, wie es in deinem Leben weitergeht* (which translates, roughly, to: "The Oracle: Paul the octopus prophesizes what will happen in your life"), whose cover features an octopus spinner that points to answers. He even prompted a full-length documentary: *The Life and Times of Paul the Psychic Octopus.*

For all of their impressive intelligence and domestic attentiveness, octopuses are, by our standards, incredibly short-lived. Even the giant Pacific octopus lives for only a few years. The male dies soon after mating, and the female begins to waste away after laying her eggs—barely sticking around to watch her thousands of delicate young float upward into the maw of a predator-filled ocean. Thus they pass along nothing more to their offspring than their genes.

Octopuses, squid, cuttlefish, and nautiluses make up the cephalopod group (cephalopoda, from the Latin for "head-foot"). And cephalopods are all members of the mollusk phylum (mollusca), making them relatives of similarly spineless snails, slugs, and oysters (*mollis* means "soft" in Latin). Octopuses themselves are an astoundingly diverse order, containing some three hundred known species. The quintessential octopus—the big-headed, thick-armed character of children's books—is actually a pretty fair characterization. In fact, there is even such a thing as the common octopus, which comes complete with a Linnaean classification: *Octopus vulgaris*. It is "chunky in appearance," according to the FAO (the UN's Food and Agriculture Organization), with what it calls "stout" (read: meaty) arms.

The common octopus also happens to be the most frequently consumed octopus. It lives practically the world over, from the cool waters off the coast of northern Spain to the warm tropics around Vietnam. It thrives mostly along continental shelves, from the tidal shallows down to more than 650 feet, and it keeps to areas of about 46 degrees Fahrenheit and warmer. But those requirements make for a pretty big territory. And with the hundreds of other species, there's an octopus for just about any water.

There's no telling what possessed the first person to pull an octopus from the ocean and call it dinner. But it was already so important to cultures thousands of years ago that it was featured on Cretan and Minoan coins (from approximately 1650 to 1500 BCE and 1450 to 1375 BCE, respectively) and inspired stirrup jar potters in the Mycenaean era (circa 1200 to 1100 BCE). Its ancient form on these vessels is slender, almost buglike, surrounded by abstract swirls of water. And since then the octopus has made it all the way to Fifth Avenue in New York City, its image scattered on these artefacts throughout the Greco-Roman galleries at the Metropolitan Museum of Art.

Octopuses, of course, far predate ancient Minoan money or Greek interest in seafood. Cephalopods have been around for some 500 million years. And the octopus's earliest recognizable forebears were skulking about the Carboniferous seas by about 300 million years ago, when the first four-legged reptiles were just starting to move up onto land. And these ancient octopuses had already evolved into something that would not look all that different from the deep-sea octopuses in the cirrina group that exists today, which have webbed arms and small fins.

The oldest described octopus fossil is known as *Pohlsepia* and now lives at the Field Museum in Chicago. The full sample looks rather more like a cow patty—flattened out into a globular splat—than a rock containing a rare, precious fossilized octopus. But researchers say they can make out its eight arms, two eyes, and maybe even an ink sac. It's estimated to have lived about 296 million years ago—even before the emergence of the dinosaurs. A 95-million-year-old fossilized octopus, uncovered in Lebanon, showed that definitely by then—the late Cretaceous, when *Tyrannosaurus rex* was terrorizing North American fauna—these mollusks had already pretty much modernized, sprouting suckers and boasting a clearly defined ink sac. "To really understand more about the ancestry of the octopus, it will be important to have earlier specimens," says Joanne Kluessendorf, director of the Weis Earth Science Museum at the University of Wisconsin–Fox Valley and one of the describers of *Pohlsepia*. But that is challenging because these animals are composed mostly of soft tissue, which does not take well to the fossilization process. Their hardest part is made of cartilage-like chitin, which does not fossilize as readily as bone.

Aristotle, one of the earliest-recorded observers of the octopus, made some surprisingly astute remarks about the animal in his fourth-century volume *History of Animals*. He was famously unimpressed with their intelligence, concluding that "the octopus is a stupid creature."

But in the intervening twenty-three hundred years, humans have gotten wiser to the octopus's different breed of smarts. The U.S. military has been funding research to study its camouflage abilities; universities are building robots inspired by its arms; and even back in the 1940s the Marshall Plan lobbed a hunk of money over to Naples, Italy, to see if a lab there could crack the code of the cephalopod brain to make more efficient computers. But despite all of the effort we have put into racking *its* brain, the octopus seems as foreign to us as ever.

Why do we have such difficulty understanding them? What sets them so apart? Maybe it's their many serpentine arms, their boneless bodies, or their otherworldly environments. But it must go beyond even these differences. We can relate to the maternal protectiveness of a deep-sea dwelling whale. We can draw familiar comparisons with a colony of insects, pointing to classes of "warriors" and "workers." But the octopus is much harder to pin down.

One thing that makes the octopus so very odd to us is its essential indifference to other octopuses. After an octopus hatches from its tiny egg and dissipates from its siblings in the ocean currents, it will live most of the rest of its life alone. Unlike squid, which often travel and feed in groups, the octopus crawls along seeking the company of its own kind only to reproduce. And unlike the committed octopus couple in the Oscar-nominated animated short film *Oktapodi*, these animals hardly establish lasting bonds. They often mate only once.

We have only recently come to recognize complex social behaviors, such as cooperation and teaching, as signs of higher cognition in other animals. The octopus's dearth of these prosocial behaviors puts it at a considerable disadvantage for gaining easy esteem from humans, at least from those of us (myself included) who have not spent a large part of our professional lives observing these cephalopods.

So to get to know and love the octopus, we must be willing to approach it with a certain amount of open-mindedness. As your humble octopus guide, I will admit to only scant encounters with octopuses for the first twenty-eight years of my life. But knowing relatively little about these creatures, I was able to discover them anew as the alien animals that they seem—to most of us—to be.

Growing up in Oklahoma in the 1980s and 1990s didn't afford encounters with real octopuses. Even the first time my parents introduced my brothers and me to fried calamari, on a family trip to the

East Coast, they refused to tell us what it was until we had sampled a few of what looked to us like mini onion rings. They weren't so bad. But even at eleven, I had to admit that I would never have tried them if I had known those light, golden brown, breaded rings contained . . . *squid*.

The only "octopus" I recall from childhood was the eponymous ride at the scrappy, now-defunct Bells Amusement Park in Tulsa. The black fiberglass ride had a menacingly painted round head surrounded by eight pivoting arms, each wielding two free-twirling cars that whipped around as the ride banked and rotated like a demonic Tilt-A-Whirl. Standing in line with my stomach-of-steel mother, we would watch the ride slowly spin to a stop and disgorge the dizzy parkgoers. You just had to keep your eye on the dripping-wet, freshly hosed-off car and hope you wouldn't get that one when your turn came to board.

It would be another fifteen years until I met another octopus, and this one would manage to actually turn my stomach. It came in the form of a seemingly harmless platter of grilled arms, served in an open-air seaside restaurant in Hvar, Croatia. I was on vacation with a good friend from college who had spent her childhood summers with aunts and uncles in Greece. A seasoned octopus eater, she assured me that the dish would be amazing. I've always prided myself on adventurous eating, perhaps fortified by the childhood calamari experience. Chocolate-covered ants at my brother's Boy Scout dinner? Done. Rocky Mountain oysters at a political convention in Denver? Down the hatch. A pile of grilled octopus arms? Should be no problem. So for the sake of the challenge, I forced down as many of the rubbery, suckery red limbs as I could and tried to forget about it.

Everything was going fine until the next morning. We had ventured to an Internet café. Down a narrow hallway was a close, stuffy little room full of other budget-conscious travelers typing away on ancient PCs. It was easy to suppress the first couple of waves of nausea. But soon there was no denying it. The octopus wanted out. I bolted for the hall but didn't quite make the front door before there it was again: on my tote, my flip-flops, the floor.

Octopus, it seemed, wasn't my thing. In fact, why would it be anyone's thing? They're slimy, spineless, solitary cephalopods that occupy the sunless crevasses below the sea. Why should we bother ourselves with them at all?

These strange animals, it turns out, have a lot to teach us—about learning, about biology, about evolution, about themselves, and about ourselves. Still, it was rather a surprise that octopuses were the reason I was standing on a dock at 5 a.m. in the dark chill of a Spanish September morning waiting for an unknown fisherman. And this is where our adventure begins.

CHAPTER ONE

To Catch an Octopus

The octopus is solitary, reclusive, and often nocturnal. Wresting one from its cozy underwater den is not a task for the weak of spirit—or the weak of stomach. Octopus fishermen often face long odds and rough seas to haul in these swarthy cephalopods. Nevertheless, this tough task is undertaken off the coast of just about every continent, via almost every imaginable method. With each octopus needing to be caught more or less on its own, the craft of octopus fishing holds a certain mystique. And to succeed, one must have a deep understanding of where the animal lives and what makes it tick. So to begin to understand the octopus, one must take to the seas.

My attempt to catch octopuses was off to a rocky start. My brand-new Canon digital SLR camera was already splattered with fish scales and puke, and it wasn't yet 7 a.m. The trip was either perfect or it couldn't have been going worse. I couldn't decide which, though, because the bucket of sloshing dead octopuses at my feet just wouldn't stay in my camera's viewfinder as the tiny Spanish fishing boat I was on heaved from port to starboard. And José Dios, the stocky, friendly captain, kept hauling up octopus traps faster than I could mumble "*Lo siento*" each time I bolted to lean over the gunwale and upchuck my nonexistent breakfast.

I had come to Spain to learn about octopus ecology from scientists in Vigo and to sample some of the local specialty *pulpo a feira*, a dish of soft-boiled octopus sprinkled with paprika, sea salt, and olive oil. But what could be better than an early morning fishing expedition with two *actual* octopus fishermen? Nothing—or so I'd thought when the idea was proposed my first morning in town. So I signed on.

Teach a Man to Octopus . . .

Fishing was probably the way humans first began encountering octopuses—these squirmy messes of lean protein. Archaeological evidence suggests that people have been catching octopuses for at least four thousand years, when ancient Egyptians waited patiently for them to crawl into submerged clay pots. It's an effective, if slow, strategy that was used from the west coast of Europe to eastern Asia and is still practiced in many places today.

The technology, however, has been upgraded a little, and the pots are not always the shapely ceramic vases that had been used for centuries but rather are often a more modern alternative. A pair of Portuguese researchers describe a new Japanese contraption this way: "The traps consist of a box, inside which a crab is tied to a string, therefore maintaining the door open. When an octopus enters the box and bites the crab, the box closes, thus preventing the octopus from escape." In many places, however, octopuses are so abundant that you wouldn't need such newfangled devices to ensure a successful catch.

In regions where they make their homes, common octopuses are around pretty much year-round. They tend to head out toward deeper waters in the winter and come in closer toward coastal shallows for breeding during the spring and summer. In high season, fishermen can collect one octopus for every three or so pots dropped. Those catches can come out to some ninety pounds of octopus per hour. It's hard work, but that haul makes for a lot of octopus salads.

Octopus fisheries experts in Spain and elsewhere object that fishing by pot (jar, box, or what have you) can be particularly hard on octopus populations. To most octopuses, a jar looks like a nice place to hide and make a temporary home. But for a female ready to lay eggs, it makes an especially nice spot to guard her brood. That means the haul includes not only a mature octopus but also all of her would-be babies—and their future would-be babies and *their* future would-be babies, and so on. Pot-fishing defenders contend that it's not so much the space that lures the octopuses into the containers but the bait inside of them. And because spawning females mostly stop eating while they tend their eggs, the Portuguese researchers note, these octopuses aren't likely to be tricked into using the traps as brooding dens. Nevertheless, to be safe, many regions have banned this method.

In the Pacific, around Hawaii, commercial fishers catch octopus via a lure on a line. The lures, now often made to look like crabs, were once simple cowrie shells. Dangled along the bottom on a long line, the lure can trick octopuses into thinking the object might be lunch. If the octopus takes the bait and grabs it, it can be impaled by several long, thin, barbless hooks and then yanked up by the fisher.

Ancient Greeks used bait to lure in octopuses: "The cuttle-fish, the octopus, and the crawfish may be caught by bait," Aristotle wrote. And the octopuses themselves could in turn be used to catch other seafood. He noted that fishermen "bake the octopus and bait their fish-baskets or weels with it, entirely, as they say, on account of its smell."

On the other hand, some octopus hunters use no lure, bait, or trap at all. Some fishers in Mauritius spear octopuses with harpoons from their boats or in the wadable waters of tidal areas. And some people even collect them with their bare hands.

Yet another more modern tactic is the bottom trawl. For this approach, a net is dragged along the seafloor, weighted down on one end to catch animals like octopuses that live on the seabed. This sort of one-fell-swoop method, however, is rather indiscriminate, scooping up anything that lies in its path. This not only can harm other animals that are caught (even if they are later thrown back), but it can also muck up delicate seafloor habitats—tumbling over rocks, killing coral, and pulling up plants. For this reason, many groups concerned with sustainability discourage trawling—and eating food caught by trawl. In Mauritian octopus trawls, for example, some 60 percent of what comes up is undesired bycatch that gets thrown back no matter what condition the animal is in. And even in the Mediterranean, where many different species, including fish, octopus, squid, and crustaceans, are kept, discarded bycatch still reaches as much as 50 percent. In Portugal, octopuses are often themselves bycatch. Those that weigh enough can be sold, but juveniles caught in the nets often don't fare so well after being thrown back. Vietnamese fishermen keep everything they haul in—whether or not it's mature—and use whatever is unsellable for animal feed, fertilizer, or fish sauce.

Trawling can also make habitats less livable for octopuses and other denizens. Some places are more vulnerable to this kind of fishing than others. Because fragile coral ecosystems can seriously suffer from a tough trawl, some areas in Europe restrict trawls to offshore muddy or sandy areas that are fifty meters or more deep. And often the octopuses out there are curled octopus (*Eledone cirrhosa*), which aren't quite as desirable as the ones closer to shore. Locals in Spain call them "the poor brother" of the common octopus caught by traps in shallower waters, so they are canned and sold abroad.

The Seven Seas

Octopuses have reached their long arms into all of the world's oceans and most of its seas. Most known species live in a thick band of tropical and temperate waters along the relatively shallow areas near the continents (convenient for us!). Some species, however, have made it into more extreme environments. The aptly named *Vulcanoctopus hydrothermalis*, for example, makes its home near the piping-hot, mineral-rich hydrothermal vents thousands of feet below the ocean surface. Others, such as the Antarctic Turquet's octopus (*Pareledone turqueti*) and Arctic "Dumbo" octopuses (*Grimpoteuthis*), inhabit deep, frigid, downright polar waters. So basically, octopuses are everywhere.

Much of the wide distribution is thanks to their long-odds strategy of reproduction. Most female octopuses lay thousands of small eggs. Upon hatching, these young'uns float up near the surface, joining clouds of other plankton for a period of weak swimming and current-driven drifting. Most of the offspring will end up as food for larger plankton feeders or as casualties of inhospitable seas. But with any luck, a few will survive the journey to adolescence and make their way down to the seafloor, where they will grow up and make thousands of their own baby octopuses.

Genetic sequencing offers the possibility at last to begin studying just how far and wide these newly hatched speck-sized cephalopods can float. But so far scientists have not yet been able to draw a definitive map of distribution. As Jaime Otero, a young fisheries biologist I met up with in Spain, points out, collecting little larval octopuses for sampling populations and abundance is a pretty inefficient endeavor. "You only obtain maybe a hundred larvae in the nets," he says. When scientists need huge numbers to have a statistically sound sample, collecting is "one of the big bottlenecks," he says.

Pulpo a Feira

To find lots of octopuses, the first place we must go is Vigo, in the northwestern region of Galicia. In fact, it would be remiss to skip this epicenter of octopuses. The city specializes in receiving, processing, and shipping octopuses from and to the world over. Huge ships pull in alongside massive warehouses in its industrial harbor, but its narrow old town streets still give off a steamy eau de octopus as the restaurants boil up fresh catch for the traditional *pulpo a feira* dish.

I nearly don't make it to Vigo. But after seventeen hours of travel from Malta, I am at last on the day's final Iberia flight from Madrid to Vigo. It is almost 11 p.m., and the countryside below is mostly black, except for intermittent clumps of light latching on to the dark floor.

As we come in for a landing in Vigo, the moon—which looks like the comforting 1920s' vintage man in the moon—has appeared in the window. A dark horizon stretches out ahead, which, I assume, must be the ocean. *The octopuses,* I think, *they're out there—all of them, cavorting about in search of crabs under the soft Spanish September moonlight*. The Ahabian octopus obsession has already taken hold.

By the light of the next morning, I see that the darkness had been not the nearby ocean but mountains wrapping the coast like a collar thrown high to fend off the stiff ocean winds. The town of Vigo is cast onto a hillside sloping down to the bay below. Otero, who has studied the local octopus fisheries, had offered to show me around.

For my first day in town, he suggests a local tour of his own devising. Little did I know that the next day I would be heading off to sea to hunt for actual octopuses. He picks me up from my hotel and we cruise along the city's waterfront, a fortress of monumental ocean-related industries. First is the fancy pleasure-boat area, where sailboats are stationed in slips and a statue of Jules Verne is perched atop four awfully octopus-like arms (despite the fact that the beasts in *20,000 Leagues Under the Sea* were squid). Next is the cruise ship terminal with its embarrassingly large floating hotel malls. Then we pass the commercial fishing section, which sends massive modern fishing ships all over the world—Africa, the Falklands, the Indian Ocean, the Antarctic—and receives the frozen catches to ship back out to the world's consumers. One big distribution company even has a giant octopus logo on its building. Next we drive by the shipbuilding area, now struggling to keep pace with countries that can manufacture fishing vessels on the cheap.

Finally, we arrive at the Spanish National Research Council's Instituto de Investigaciones Marinas, where Otero had done his octopus research. The building is a low-slung mid-twentieth-century number. There we meet Ángel Guerra, a well-groomed older man in a plaid button-down shirt. "My old boss," Otero introduces him. After walking us

Jules Verne sculpture in Vigo. (Katherine Harmon Courage)

upstairs to his tidy office, Guerra pulls up two decades-old rolling chairs and asks me what, exactly, it is that I have come to learn about octopuses.

This is a question I was getting used to hearing after announcing to scientists, who had spent decades studying cephalopods, that I, with but a bachelor's in English, a master's in journalism, and a magazine job in New York City, was writing about octopuses. I explain to him that I am there to learn everything I can. But it cannot possibly be boiled down, he says. As a civil servant, he has been working for the Instituto for nearly four decades. He started when there were just a handful of others, but now there are more than two hundred people there studying cephalopod ecology, fisheries, and oceanography as well as marine food technology. I have come to the right place, he says, but my timing is bad. You see, in Spain they have at least two or three octopus feasts a month between April and September, yet I'd managed to arrive just a couple of weeks after the last big one of the year. I must stay for at least a few weeks to learn about the octopus and its place in the culture, he says. Alas, I cannot. So he gives me a sixty-minute summary, listing his research highlights and telling tales of doing the *tako-tako* octopus dances he'd learned in northern Japan. Almost midthought, he glances at his watch and suggests that it is time for a coffee break.

We head out the front door, across the main road, and down a small street lined with cafés. At one, a boisterous group of young people is sitting out in the sun enjoying small coffees, slices of bready cakes, and Spanish tortillas with egg and potato. This, it turns out, is Guerra's lab team. As soon as I shake hands with all the men and trade cheek kisses with the women, Ángel González, a tall man with curly dark Spanish hair, is peppering me with stories of his time studying in the States and his trips to New York City.

Eventually, the coffee consumed and the snacks eaten, the group reluctantly gets up to return to the lab. But first, Guerra says, we must swing through a local fish market. It is already emptied of octopus for the day, but it still has on display loads of shining fish, crabs, and squid on ice—all of which could easily sneak up on you because they have absolutely *no* smell. Not even a hint of the sea's brininess. After having walked through the streets of New York City's Chinatown so many times in the hot summer—or even the frozen middle of winter—it is a revelation that a fish counter could be totally free of that pungent odor. It makes seafood—a food group that categorically repulsed me as a child—seem a much more natural and approachable staple.

When we arrive back at the labs, Guerra takes me on a quick tour of the building, a labyrinth of offices and laboratories. We poke our heads into one room where a team has banks of aquariums in which they are trying to rear delicate sea horses to repopulate areas where they have been depleted. Next we arrive in an office where González is talking intently into a telephone receiver. *"Sí, sí, mañana si posible,"* he says, after which I lose his quickening stream of Spanish. After what I gather is a bit of explanation about my research and quite a bit of cajoling, he scribbles some notes on a Post-it and thanks the other party profusely.

I am going on a fishing trip tomorrow, it appears. He hands Otero the Post-it and says we just need to get an insurance policy from a particular agent so that I am covered for my trip to sea—or else I cannot get on the boat. Insurance seems excessive and a bit ominous, but I'm so excited about the possibility of the expedition that I don't care.

Otero and I set out to find this insurance agent. And after much negotiating and waiting and more negotiating, we manage to secure my necessary paperwork for the trip. "Guard that like gold," Otero says of my insurance packet. "Of course! It's my passport to the sea," I say lamely, stuffing it into my tote bag. And we head off to find some *pulpo a feira* for a late Spanish lunch.

We walk down Vigo's tourist strip, named for the *pulpeiras*. In this traditional octopus preparation method, women, the *pulpeiras*, boil the day's octopus catch in big barrels right on the street. Off the main drag, in an old square, we step into a small restaurant with half a dozen wooden tables. Otero orders us a maritime feast and explains each dish as it arrives, carefully writing down the names for me in my notebook, being sure to include the Latin names of each species we consume. Everything is fantastic—fresh, light, and delicious—and the local Albariño white wine, with its slight bite of acidity, is perfect. But the main feature is the *pulpo a feira* (or *polbo a feira* in Galician), which means something like "fair-style octopus." It is exquisitely tender. The skin layer has an almost fatty texture, which is artfully balanced by the granular sea salt.

Pulpo a Feira

Courtesy of Jaime Otero

Defrost an octopus, which has ideally been frozen for one to three days—a process that breaks down the tough fibers of the muscles, making them more tender.

Boil it in water and olive oil for about 30 to 40 minutes (or until almost tender, depending on the size of the octopus).

Optionally, add pieces of potato to the boiling water for the last 15 minutes of cooking.

Remove the octopus and potatoes from the water when they are both tender.

Cut the octopus arms into small pieces with kitchen scissors.

Serve octopus and potatoes on a plate with olive oil, sweet and spicy paprika, and salt.

Upwelling

Vigo is perched on the fertile coastal edge of Galicia in northwestern Spain. The area is dominated by rivers that flow out to sea in five separate estuaries or embayments—locally known as the *rías*. The surrounding coast is rock and, in many places, quite steep. Otero and I drive along the smooth, curvy roads. When we catch glimpses of coves and dropping cliffs, Otero points out his favorite spearfishing spots. The craggy rock continues below the surface, providing perfect hideouts for octopuses and their favorite local food: blue crabs.

A dense fog is rolling in, and eucalyptus trees hang on to the steep embankments, making the whole scene indistinguishable from northern California—complete with the vineyards and excellent wine. It's actually a very similar climatic environment, Otero points out. The cliffs, the flora, and even the fog are all part of a system that starts out at sea with the ocean and wind currents. It is these winds, currents, and cliffs that have made places like Galicia an Eden for octopuses—and, in recent centuries, also for the people who fish for them.

The oceanic process behind all of this is called upwelling. It occurs when cold, nutrient-rich waters flow up from the deep and move into the shallower areas closer to the shore. This upwelling brings with it a feast for young octopuses in the area. As the seasons and the prevailing winds change, so does the underwater current, eventually starting a downwelling cycle. But between these welling events are brief seasons of relative calm, when wind and currents relax, not pushing or pulling in their normal frantic fashion. The transitions keep new hatchling octopuses close to the coast, protecting them from being swept out to the deeper, less hospitable seas.

We arrive in the small fishing town of Cangas do Morrazo, its harbor full of small colorful boats. These modest craft are somehow responsible for the bulk of the Galician catch, which is no small feat. It is on a boat like this that I will be venturing into octopus territory early the next morning.

Along the harbor edges, piles of what look like homemade lobster traps are stacked neatly. These traps are crafted from various combinations of wood, carefully welded metal, tire rubber, plastic pipes, and netting. And they are not, it turns out, made to catch lobster. Rather, they are creels made for catching other bottom-dwelling, benthic beasts: octopuses. Each is connected along a long line, which is unfurled, creel by creel, to rest on the bottom. There they sit, waiting for an octopus or two to crawl in—lured by a pouch of stinky seafood bait. The beauty of the design is that the octopus could find its way out if it wanted. "But inside they have food," Guerra had explained that morning. From the octopus's perspective, he concludes: "So why would I want to go out? I am very comfortable in here."

The sea's harvests are brought ashore and sold right away at tiny dedicated buildings called *lonxa,* which means auction in the local

Creels stacked on the docks in Galicia. (Katherine Harmon Courage)

Galician dialect. These are run by the local fishing cooperatives. In the old *lonxa*, the day's catches are piled by type on the floor, and a caller sings out the prices in Galician as local buyers for restaurants and shops place their bids. Traditionally, Guerra says, the women would be the ones to purchase the day's goods. The caller starts at a high price and counts down (an interesting reversal, perhaps reflecting an assumed high inherent worth of the catch). When a price seems right, a buyer yells out, "*Hep!*" Stop!

Of the several dozen *lonxa* in Galicia, the old stone buildings are slowly being replaced by modern auction houses, where the goods move down a conveyor belt in trays. A screen ticks down the prices and buyers simply have to press a button to make their purchase. Otero seems to think the new auctions will bring more efficiency to the process—or at least there will be "no spitting," he jokes. But that also means that no longer will locals be able to buy their seafood for a song.

For his research, Otero had been interested in assessing just how much octopus the local artisanal fisheries were actually bringing in. Some of his papers had included interviews with local fishermen about their octopus catches and fishing habits. Only now was I noticing that Otero—with his Converse sneakers, hipster T-shirt, and Ray-Ban glasses—though from the region, seemed almost as out of place on these working docks as I did. He would not exactly chat up the octopus fishermen. "You don't want to mess with fishermen," Otero says turning toward me. "They always carry knives." A local technician who knew some of the local fishermen had done the interviewing, he says. "I just analyzed the data." The next day's trip was suddenly sounding a little more foreboding.

In spite of their toughness, Spanish octopus fishermen, I learned, are hardly different from American bass anglers or even Okie noodlers. Most of them have their favorite lucky spots, whose locations are kept as carefully guarded secrets. And with more than nine hundred miles of craggy coast in Galicia, there are plenty of secret spots.

In Spain each fishing boat has a logbook in which fishermen are required to record their catches. But as Otero explains, a bit of fatigue crawling into his otherwise energetic voice, often the fishermen "just don't care." In Galicia alone, there are some four thousand of these small vessels that bring in the majority of the region's catch, "which means that the small-scale fishery in Galicia is huge and has a very high social importance," he says. So it's kind of a touchy subject when it comes to proposing new regulations. Especially with so many knife-wielding fishermen.

The Spanish researchers also have no way of knowing how much effort the octopus fishermen are putting in to catch what they do. Are they pulling up mostly full creels and only fishing for part of the day? Or are they casting the traps out like crazy all day only to reel in a few per line of creels? These questions are not necessarily out of concern for the fishermen's overall job satisfaction or general morale, but rather they are important for measuring the health of the local octopus population.

We looked out over the harbor of colorful, bobbing boats. With each tiny boat being manned by two or three fishermen pulling up creel after single creel, and tubs of octopuses being dumped onto the floors of local auction houses to be sold literally by a Galician song, the whole thing suddenly seems impossible to track. "Does it give you a headache thinking about trying to manage a fishery like this?" I ask him. He pauses and exhales. "It is not for me," he says. I ask whether it could be done at all. "Maybe it's not possible."

Our final stop of the day is Bueu, the town from which I'll be leaving on my expedition early the next morning. We walk down to the harbor to find the boat where I am to meet my local fishermen at 5 a.m. Walking down the second dock in Bueu's harbor, as we had been instructed, we hunt for the name of the boat. And there she is.

The *Nuevo Carolay,* with an attractive blue deck to offset the white pilothouse and strip of red roof, looks like most of the other fishing boats tied up in Bueu's harbor. As a bit of whimsy, a stuffed toy cow is tied up to the light beacon on top. Shrunk down, she would have made a perfect bathtub toy. Little did I know that despite the clear skies, this ride would be more like the bath of a tempestuous toddler.

We stop briefly at a local bar before heading back to Vigo. It's just getting dark, and I'm surprised to see that it's already 9 p.m. The September sun sets late on the northwestern corner of the continent. Otero

The *Nuevo Carolay* in Bueu's harbor. (Katherine Harmon Courage)

confesses that he's "brainwashed" and ready to hit the sack for a few hours before we have to drive back to Bueu in the darkness of the Spanish morning so that I can hunt octopuses.

Pescadores

It's pitch dark at 4:45 a.m., and I'm shivering on the pier, feeling both under- and overdressed in my khaki pants, embarrassingly nautical navy-and-cream-striped top, light cotton sweater, and running shoes. Otero is standing next to me in a light jacket, scanning the dark docks for our fishermen.

The harbor is quiet and still full of fishing boats. Men slowly start arriving at the harbor by ones and twos, most wearing track pants and sweatshirts, carrying small tackle boxes or plastic bags, which must be lunch. Two fishermen signal to each other from across the dock with mock seagull calls, an imitation so perfect that it could only have come from prolonged proximity. At last, a short, sturdy but friendly looking guy approaches us (how out of place we must look, slender kids standing with arms crossed against the chill).

This fisherman is José Dios, the captain of the *Nuevo Carolay*. "*Los dos*?" he asks. No, Otero explains. It's just going to be me. "Is she going to be okay?" the fisherman asks in Spanish, gesturing at my pants and shoes. I'm starting to wonder.

Dios takes us down the dock to the *Nuevo Carolay*. As soon as we come alongside the boat, I'm surprised to see a diminutive older man in yellow rubber pants and jacket already working in the back of the boat. How did he sneak in without our noticing? He looks up at us from the dark, a bright light from the pier shining into his face. He looks suddenly tough, like a startled creature that has learned not to show fear.

Something is explained in Spanish, and Dios helps me down into the boat. It's so small that I don't even know where to stand. So I stay where I am, my back against the tiny pilothouse as Otero and Dios exchange a few words. The fishermen are planning to stop at an island for lunch. From there I have the option of catching a ferry back to

Bueu—in case I am tired or seasick. "Just in case," Otero repeats for me. Okay, I say, wondering why I would want to do that.

I can tell, though, that Otero and I are both wondering how—*whether*—I will manage in this small floating world of knives, galoshes, rubber overalls, squirming sea creatures, and meaty-forearmed Galician men with my thin sweater, fancy sneakers, and fragmented South American Spanish vocabulary. Standing there in the dark chill of 5 a.m. on a working fishing boat, clutching my canvas tote, it is all starting to seem like a crazy and possibly dangerous idea. But at this point, there is really no other choice.

By 5:15 a.m., we are at last slowly pulling out of the slip. Dios has donned a baseball cap and red rubber overalls over his gray, hooded cutoff sweatshirt, making him look like a rugged midwestern farmer or football coach. He cranes his head out of the side window of the pilothouse to see where to steer, just as one might back a truck out of a parking space. Otero walks back up the dock to his car to drive off in the darkness on the smooth, solid highway toward sleep and civilization in Vigo.

At my feet, big bins of smelly dead sardines stare back at me with blank white eyes. The other man on the boat, Manuel Manolo, has switched on a hose and is spraying water over the crates of fish. Somehow Dios conveys to me that I should step over the high sill and join him inside the pilothouse, in which there is barely enough room for the two of us to stand. The dash is covered by monitors, a panel of switches, a roll of paper towels, and an old beach towel. Dios continues to navigate with his head out of the side window. The front windshield is too covered in saltwater spray to see much of anything but darkness and occasional distant blinking lights. I try to put together some friendly conversation: *"Es muy calm, uh, calmo?"* I venture. *"Ah, sí, calma,"* he says graciously, adding something about *tardes*. Perhaps it was not so calm yesterday afternoon, I think, optimistically, not venturing to consider the future tense.

The little boat chugs along. Dios has switched off the top light, leaving us a dark ship pushing on into the night. I try not to wonder whether that is safe and instead focus on the almost comforting rhythm of mechanical noise put out by the engine and the chatter and static coming out of the CB radios that are mounted just above Dios's head. One voice comes through as a dead ringer for a character from Alexander Korda's 1930s rough-and-tumble Marseille sailor film *Marius*. No matter that the language, country, and century are different—the sea is the same. The coastal lights are fading into the distance; the moon above and a faraway beacon become the only points of light. The even darker realm below takes on a new density.

Once we hit more open waters, Dios kicks the speed up to twenty knots. He reaches across me to close the side window just before the spray can make it inside. We are headed toward a lighthouse and a row of scattered lights, which turn out to be other small fishing boats, already at work pulling octopuses up from their watery world.

As we get closer to a rocky shore, Dios slows down, pushes a few buttons to activate one of the monitors, and an incomprehensible smattering of lines and dots appears. With a few beeps, he seems to have found what he is looking for. He cuts the engine and steps over the sill to the port side of the pilothouse, where there is a hip-high metal trough that runs down the side of the boat. Manolo is already standing midgunwale ready and waiting.

A creel appears over the side of the boat, and then another and another. Dios is pulling creel after creel up and onto the gunwale. Most of the first traps contain a small zoo of crabs, starfish, or flopping orange and silver fish. But this is all detritus for an octopus fisherman, so Dios dumps the stowaways back into the dark sea below. He then grabs a couple of sardines from a nearby tray and with a distinct *skush*—audible even over the idling engine—he squishes them into a plastic mesh bait pouch among old fish bits from the previous day. Manolo finishes preparing the creels for the next deployment and then walks—with an almost cowboy-wide gait—back to the end of the boat to stack them in neat rows.

All of a sudden, the mute Manolo becomes animated and excitedly says something in my direction. When I turn to look, Dios seems to be doing battle with an incredible sea monster that is trying to wrap its

Early catch of the morning with the *pescadores*. (Katherine Harmon Courage)

flailing purple limbs around his bare arms and onto his torso. From my vantage point between the pilothouse and the crates of sardines, I can hardly see around the side to where Dios is working. But I hear a ripping sound, and with a flick of his left arm, Dios holds the now limp and pale octopus around the corner for me to see. I'm so stunned that I can't even bring my camera up to my eye before *slop*, the octopus is flung into a big bucket at my feet.

More octopuses come up, and each time, with a big grin, Manolo points to make sure I see it. All I can think to do in my disorientation to convey my excitement and appreciation is exclaim, "*Oy! oy!*" The biggest ones are probably five and a half or six pounds. Dios—suddenly the serious one—holds these hefty specimens sucker side up, their arms writhing in the cool dark air, around the corner of the pilothouse for me to photograph before he slays them and tosses them in the sloshing bucket that is quickly filling up with dead octopuses and seawater.

All of the animals are a deep midnight purple when they are yanked from the cozy creels. Some of them flash a patchy-brown rocky color as they're removed, and most, as they are dispatched and tossed, hardly moving, into the tub, have faded to a ghostly Goya white, the color of a nightmarish fever dream of flavorless deep-sea kraken.

Some of them are still squirming in the bucket. And one white arm has reached out above the rim in a futile last effort of escape, where it then hangs, suctioned on but lifeless. Eventually they settle into a palette of pale purples, browns, and grays, molted with whitish spots. Some of their eyes peek out amid the swirl of arms and old-man-hands blotched mantles. They look neither tasty nor intelligent. Maybe it's the fatigue or the cold or the lack of breakfast, but I feel no remorse, no pang of sadness, as these animals are dredged up from their rocky world a dozen meters down and flung into a tub of death. These creatures seem simply like

Bucket of octopus catch. (Katherine Harmon Courage)

dumb wild animals, all arms and droopy flesh. Just a messy, feisty catch, far less elegant than the sleek fish that sometimes sneak in among them.

The octopuses seem to come up in clumps—sometimes even two to a creel. There would be a dozen empty creels (clogged with crabs and starfish) and then *uno, dos, tres* creels would come up bearing cephalopodian prizes. *It must be kind of exciting,* I think, standing there at the winch as Dios is, while each creel comes up from the darkness until it is just visible beneath the surface, holding a possible present dislodged from the dense pull of the sea, each octopus a sixteen-euro bundle of flesh, a gift from Nature herself.

The small ones that are about two pounds or less—*los pequeños*—get tossed back overboard, per regulation. I can see their pale webs outstretched as they drift slowly downward into the darkness. I try to keep count of the keeper octopuses as they *sploosh* into the container at my feet. But I soon lose track, and the creels keep coming: pull, open, empty, bait, slide, bait, close, slide, stack. The almost mechanical process—surely a tiresome task for Dios and Manolo—is stabilizing as I strive to keep my balance in the rolling boat.

At last a blue jug comes up on the rope instead of a creel: the end of the line. Dios dunks his hands into a bucket of seawater, rinses off the slimy front of his rubber overalls, and then steers the boat back out away from the shore. Suddenly I register a gentle, measured *splish . . . splish . . . splish.* I look back to see Manolo tossing the creels back into the water in perfectly measured increments. The rope holding them together unfurls up and down the length of the boat and then overboard with the next creel at a dangerous pace. Suddenly he shouts: *"Ho!"* The last creel pops overboard, and Dios kicks up the motor. After washing down the deck, Manolo stands at the back of the boat meditatively and lights a cigarette in the dark. We motor off to the next secret octopus fishing grounds.

Rough-and-tumble Seas

It's not quite 7 a.m., and it's still dark in Galicia. I've lost track of where we are in relation to Vigo. The boat slows as we pull in perilously close to some rocky cliffs. Manolo is standing on the port side, holding a broomstick topped with a menacing-looking metal hook. He gives a cry and Dios stops the boat. Manolo snags the next blue jug, marking another set of creels that had been laid the day before. Without another word, the two men go to work bringing up the next strand of starfish, crabs, and the occasional octopod.

One slippery octopus slides right out into the metal trough, attempting to slink under the creel as if it might sneak away unnoticed, but it is not quick enough. Dios snaps it up and dispatches it with his knife—into the bucket with the rest of 'em.

Another creel comes up heavy and full of a mysterious black mass. Dios opens the creel and unfurls an impossibly long, thick, and shiny eel. It flops onto the deck and wriggles with surprising ease among the ropes at the fishermen's feet. The two tiny, frilly front fins give it away as a conger eel. It's purple-black with a dark dorsal fin extending down its whole body and a creepy white underbelly. With its tooth-filled mouth flared open, it looks sea monster–like, a mythical thing of marginalia and antiquarian maps—and ready to attack. After it starts flapping its long tail against Dios's rubberized pant legs, Manolo bends down and tosses the beast back overboard.

So this is the octopus's world: a feast of tiny orange crabs, schools of slick silver fish, weedlike starfish, and terrifying prehistoric sea

Conger eel loose onboard the boat. (Katherine Harmon Courage)

serpents. It's hardly a scientific sampling, but the bycatch starts to paint a picture of the hidden submarine faunal landscape.

The eel is one of the octopus's great nemeses. Moray eels prey on octopuses and even seem able to sense their presence via chemical signals. Being lean, these eels can often make their way through even a small opening into an octopus den. Who eats whom, however, can come down to a question of strength, because an octopus will devour an eel if the eel doesn't get the octopus first. This slippery wrestling match has played out right in front of scientists' eyes. At the Scripps Institute, researchers placed a medium-sized moray eel into a tank with a two-spot octopus, Frank Lane noted in his classic 1960 book, *Kingdom of the Octopus*. "The octopus attacked the eel repeatedly. For a time the eel's agility enabled it to escape but eventually it tired and the octopus killed it and ate part of its head." Things don't always work out so swimmingly for the octopus, however. As Lane describes, it also can be that

> [t]he eel applies a different technique—as gruesome an example of animal mayhem as can be found in all of Nature. Grasping an arm in its vice-like jaws, the eel stretches its body full length and then spins itself round and round until the arm is twisted off. It is then swallowed, and unless the octopus has managed to escape during this slight pause, the eel continues to eat it an arm at a time. The spinning technique requires only enough space for the eels' straightened body, and an octopus caught in its lair can, therefore, be eaten piecemeal with little possibility of escape.

In addition to the eel, the octopus faces plenty of predators. A full-sized octopus can be food for barracuda, grouper, yellowtail snapper, squirrelfish, lingcod, scorpion fish, sharks, sea otters, seals, birds, killer whales, and many other hunters. Not all octopuses, however, will allow themselves to become meals. Considerable size and cunning are needed to take out the giant Pacific octopus, which has been known to kill dogfish sharks. (A video from the Seattle Aquarium's massive four-hundred-thousand-gallon tank shows a sixty-pound male giant Pacific octopus taking out a three-foot-long dogfish shark, strangling it with its eight powerful arms.)

When it can't take out a potential predator with sheer strength—or would simply prefer not to be bothered by other less threatening animals—the octopus has developed an extensive arsenal of other defense mechanisms. If they sense danger, most octopuses can shoot out ink (deep-sea species are the main exception, as their world is so dark that a cloud of ink would make little difference to a would-be attacker). The ink functions like a smoke screen to obscure an escape. In addition to clouding the vision of an attacker, the ink contains a compound called tyrosinase, which is thought to irritate an assailant's eyes and confuse its smell and taste, throwing it off the trail.* But if an octopus isn't able to escape its own ink cloud, it can die—which makes shipping live octopuses a chancy endeavor, with a risk they will inadvertently poison themselves. In some cases, the octopus can mix the ink with thick mucus, which creates a dark viscous blob that hangs in the water as a pseudomorph to serve as a decoy for a predator. Meanwhile, the octopus itself can turn a different color and try to jet away from the distracted attacker. This thicker ink can also do an extra number on a predator by jamming up a fish's gills.

Some species of octopus have even been rumored to pick up weapons for defense. The Atlantic violet blanket octopus (*Tremoctopus violaceus*) has been observed wielding pieces of the toxic tentacles of the Portuguese man-of-wars (*Physalia*). These octopuses are apparently immune to the sting and likely snag them straight from the man-of-wars and clutch them with their suckers to deter predators.

So impressive are the octopus's adaptable evasion tactics that University of Arizona marine ecologist Rafe Sagarin proposed them as the best animal for us to study for our own defense efforts, as he describes

* It doesn't take much ink to put off a human handler, either. As Woods Hole biologist Lydia Mäthger and plenty of other cephalopod researchers have figured out through experience, if you're going to irritate an octopus, it's best to wear old, dark-colored clothes—or better yet, a trash bag over your garments—because, as she notes, "It destroys the clothes." Even if you rush to wash off the ink right away, it won't come out. "Forget about it," she says.

in his book, *Learning from the Octopus: How Secrets from Nature Can Help Us Fight Terrorist Attacks, Natural Disasters, and Disease.*

The octopus's primary security strategy, however, is not all that novel or courageous—and it makes 'em gosh darn difficult for us humans to catch. These soft-bodied animals spend much of their time hiding out in the relative safety of their dens.

Humans, of course, have figured out ways to lure these wary animals out of their lairs—namely with stinky bait and inviting traps.

As the fishermen in Galicia are reeling them in, one by one, a faint glow begins to seep into the morning sky. Some patches of green vegetation clinging to the granite rocks become visible, as do the waves crashing against the nearby cliffs. A sliver of light catches the gulls and surface of the sea. I realize deeply, as if in a dream, that this place, wherever it is, is one of the most ruggedly beautiful places I have ever been, the type of place that will let you stay just long enough to see its exquisite savageness. As we motor away, Dios mentions that his family has been coming to this spot for years. "*Bonita*," he says, almost to himself.

The next leg of the trip takes us out into choppier waters. The boat is pitching wildly despite the clear skies. I've wedged myself into my corner of the pilothouse, holding on to the top of the doorjamb and pressing my right arm against some metal protrusion on the side, trying to act casual.

The next fishing spot is hardly any more placid. Dios steps out of the pilothouse as steadily as can be, and starts pulling up the next batch of creels without so much as a stagger. I've climbed out and am now bracing myself on the other side of the doorjamb, trying to stand steady enough to take a picture of the seagulls, which are suddenly swarming the boat. They are crying and calling, some almost within arm's reach. They beat their wings furiously to stay in place in the punishing morning wind, some flying sideways while others seem even to be flying backward. I keep an eye out for any hint of bloodlust, but they appear to be interested only in the fishy detritus that's being cast back to sea from the creels devoid of octopuses.

The three hours of sleep from the night before seem days in the

Manolo stacking creels in the stern of the boat at dawn. (Katherine Harmon Courage)

past. Three long, windy, choppy days ago. I try to focus on the cool wind, the beautiful cliffs, the creepy gulls. But I close my eyes for just a few moments, and then, suddenly, I start to feel it. I know what it is, but I try to tell my stomach, my esophagus, my inner ear that everything is fine. *This is nothing. Just a passing wave—ugh, don't say wave—of, of nothing.* The gulls are still lurking above like a scene from Hitchcock, and Dios and Manolo are hard at work, unaware, I hope, of the drowsy writer poised above the caldron of dead octopuses that are slowly swirling with the heaves of the boat. *Just focus on the fresh air and perhaps a picture of the gulls,* I tell myself. *Or maybe I can just suppress it. No one will have to know . . . No, no, here it comes.* I stagger as close to the starboard side as I can get, crushing my tote bag and camera against the gunwale. I puke onto the arm of my white sweater, with some remnants on the railing, tote, and, of course, the camera.

I bolt for the roll of paper towels in the pilothouse and am just wiping the puke from my chin when Dios steps in after washing down from this haul. *"Mal?"* he says. *"Sí, un poco,"* I say apologetically. He shrugs kindly as if it's no big deal. It's *"naturale,"* he says. I push up my puke-stained sleeve despite the cold. *"Mejor?"* he asks. *"Sí, mejor,"* I say, because I do actually feel better. *"Dos mas,"* he says, holding up his thumb and index finger. Just two more stops. *No problemo!* Now that that's over, I'm ready to, er, roll.

We head back out, and the waves are even bigger now, pitching the boat up over the crests and then letting it wriggle into the valleys. But the sky is a clear blue, and it's beautiful on the fresh open water. The stop out here turns up new species in the creels. Some silver fish with unreasonably large eyes flop around in the metal trough. Manolo hands one to me to feel. It's perfectly smooth, and its delicate scales come off on my hands. As Dios displays another octopus for me, I toss the fish overboard and grab for my camera. He brings this one around for me to see live. He flips it over, and without explanation, he grabs a long white plastic tool and stabs the thing right in the mouth. *Right in the mouth!* With a flick of the wrist, he turns the now-limp octopus over before tossing it in the bucket. The seas are a rough place indeed.

After one more octopus stop and another *naturale* episode over the side of the boat, it's time to break for lunch. That ferry back to Bueu is sounding better by the minute and lunch absolutely impossible. Before

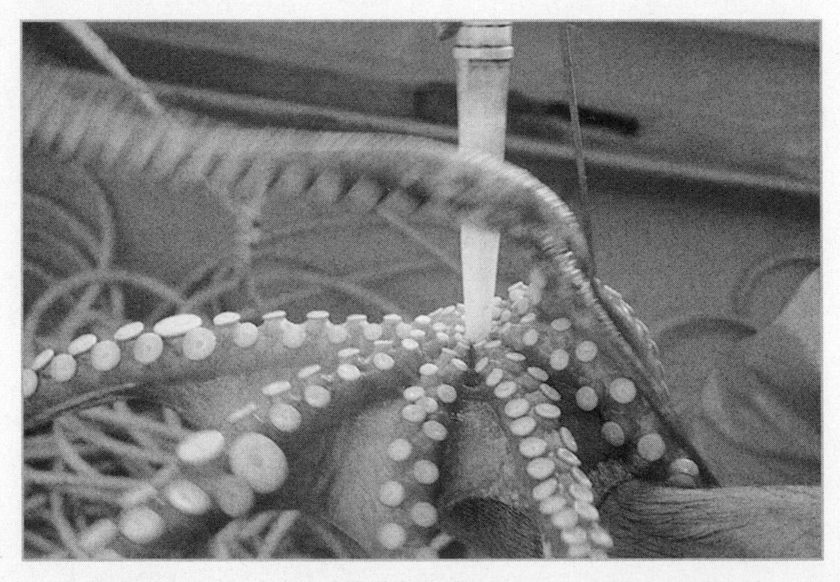

Dios demonstrates how to kill an octopus. (Katherine Harmon Courage)

disembarking, I ask Dios how the haul has been. "*Bien,*" he says with the cautious modesty of a fisherman not wanting to jinx his good luck. So far that day, Dios figured, they had brought in a little more than 150 pounds—maybe sixty-five from the first stop alone. This year has been not quite as good as the last. But it's like that, he says, sometimes it's good and you get a lot, and others, not so much. These are the warm, quiet days when the winds are just shifting from the northerly gusts of summer to the southerly winds of winter. What must it be like, facing far rougher waters, during these crapshoot creel tosses day in and day out? Apparently it imbues placidity—if it doesn't breed mania.

One hundred and fifty pounds sounds like a whole lot of octopus for a morning's work. But if, as Otero had said, an octopus weighing about four pounds was going for about two euros per pound at the auctions, that means, roughly, a daily gross of three hundred euros for the two of them, minus fuel, equipment, and boat maintenance costs, and local fishing cooperative dues. It might have been the fatigue or

Manolo (left) and Dios with two of their biggest catches of the morning. (Katherine Harmon Courage)

the residual nausea, but for all the hours these guys—and the thousands like them—spend reeling octopuses up from the rocky Spanish ocean floor, they started to seem almost heroic.

A Global Catch

Octopus is no tuna, no cod, and certainly no salmon. But even if *you* don't often encounter it on your table, millions of people around the world do. In fact, as you read this, *right now* thousands of *tons* of already-caught octopuses are moving around the globe on the high seas, the highways, and even flying through the air. They really are everywhere.

Although octopus has been a popular dish across Asia, the Mediterranean, and other regions for millennia, not all parts of the so-called civilized world have been on board with eating octopus. Back in 1963, for example, biologist and fellow of the Royal Society Malcolm R. Clarke humbly proposed in *New Scientist* magazine that with the octopus's "high protein content, 20 per cent, which compares favorably with economically important fish . . . [i]t could, in principle, serve as human or animal food." He does submit, apparently from a gustatorily disinclined perspective, that some people might be a little skittish about such strange creatures creeping onto their plates:

> It seems doubtful if cephalopods will ever be used extensively for human food in Britain; even on the continent, where familiarity should have overcome any prejudice. However, there is no reason why cephalopod should not be used for animal food, bait or manure—as it is in other parts of the world.

In the intervening half century, however, this high-protein food has become increasingly popular. During the past fifteen years, Galician fishermen alone have brought in an average of three thousand (reported) tons of common octopus each year. And although they certainly do a valiant job of eating loads of it locally, most of it gets shipped overseas.

But as with the dynamics of much modern commerce, distribution does not always appear logical at first blush. In 2000, for example, according to fisheries records, the Spanish were exporting 27,190 tons of common octopus—but *importing* some 30,404 tons of the stuff for their own consumption. The global octopus trade seems to go like this. Even regions that are rich in octopus often sell it to countries that will pay more and buy their own octopus from countries where it is cheaper (often owing to low labor costs and less regulation). Japan, for example, though it has its own impressive octopus population, now gets much of its supply from North African countries.

In the late twentieth century, Japan saw about a 50 percent drop in its local catch of octopus between the 1960s and 1980s alone. With the decline of Japanese stocks, the country launched a long-term project to improve the octopus spawning areas in hopes of boosting populations. They dropped extra rocks and some twelve thousand to seventeen thousand brooding pots each year to provide uncollected shelters and egg refuges, making the area much more habitable for octopuses.

North America doesn't have such a long-standing octopus-fishing tradition, but there are a few legitimate fisheries on the continent. In Mexico's Gulf of California local fishermen catch two-spot octopus—*pulpo manchado* (*Octopus bimaculatus*)—in the winter and Hubbs's octopus (*Octopus hubbsorum*) and Pacific red octopus—*pulpo rojo* (*Octopus rubescens*)—in the summer. Teams of two or three fishermen set out in one of that region's fifteen hundred or so small boats called *pangas*. Many teams dive for octopus using long compressed-air hoses rather than expensive scuba equipment. While one member dives down with the air hose to catch octopuses by hand, another stays on board to man the *cabo de vida*, keeping tabs on the hose and air compressor. In shallower areas off the coast of northern Jalisco, Nayarit, Sinaloa, and Sonora, divers swim down and snag octopuses by gaff (a pole with a hook or barbed spear). And in Baja California Sur, fishermen use traps in the rocky areas, casting five to fifty traps down to the floor, anywhere from 6 feet to 165 feet below. These methods make for a considerably more sustainable alternative to the larger trawl operations elsewhere. But most people don't bring in much more than a hundred pounds per day, according to fisheries reports.

Out in Hawaii, the most commonly caught octopus is—you guessed it—the Hawaiian octopus, also called Cyane's octopus (*Octopus cyanea*), which is found in other warm waters throughout the Pacific and Indian oceans. Around Hawaii, it's sometimes referred to as the big blue octopus—owing to its default color—or as *he'e mali*, *tako*, or even, strangely, "squid." Only some 11.7 tons were caught commercially in 2000, mostly by spear and jigging with a lure. On the islands, it is often served in the raw Japanese *tako* style, as *tako poke* (diced in a vegetable salad), dried, frozen, or smoked.

The continental United States has its own octopus population, including some common octopus (which might actually be *Octopus americanos* or *Octopus insulares*, depending on whom you ask), which lives along the southeast coast, and the giant Pacific octopus, which lives in the Pacific Northwest. But none of these populations has yielded much of a formal fishery. Instead, the United States gets much of its octopus from the Philippines—about 3,750 tons annually. More than a dozen shallow-water octopus species live in the Philippines (including *Cistopus indicus*, *O. aculaetus*, *O. cyanea*, *O. luteus*, *O. nocturnus*, and *O. ornatus*), but as noted by Seafood Watch, an organization at the Monterey Bay Aquarium that makes science-based recommendations for sustainable seafood choices, none of these populations is managed or has even been properly assessed. And tracking individual species is nearly impossible because consumer products are generally just labeled as "octopus."

Most of the sushi-grade octopus in the United States is *Octopus vulgaris* and comes from Mauritania, Morocco, Spain, and Vietnam. But it often makes a stop in Japan first to be prepared. This processing stopover makes it difficult to figure out where a piece of octopus arm originally came from, because it will usually arrive in the United States simply marked as a "product of Japan." Vietnam, for example, exported more than twenty-three thousand metric tons of octopus in 2004—the largest share of which went to Japan.

In general, Seafood Watch suggests that consumers avoid ordering *tako* sushi simply because it's nearly impossible to tell where (and how) it was actually caught. It's even difficult to tell what species it is. And based just on some supertough takeout *tako* I've tried, I would recommend avoiding any but the best unless you're looking for some serious jaw exercises.

Morocco, Mauritania, Senegal, and other coastal countries in Northern Africa are now some of the major suppliers of common octopus to the world. But that wasn't always the case, Ángel Guerra says. He did some of his doctoral research on octopuses in northwestern Africa, which had once been part of the Spanish empire. Although the primary focus had long been fish, while he was there in the 1960s the prize began to shift to cephalopods, he recalls. "It's a matter of overfishing, and it's a matter of oceanographic conditions that have changed. And the octopuses—and the cuttlefish and the squid—are much more successful than the fish." That success made them the next targets of commercial fishing.

Morocco, in particular, became a serious octopus supplier starting in the 1960s, as their other fish stocks declined. But fished by local Moroccans, as well as Spanish and South Koreans, octopus populations there took a hit starting by the 1980s. Some subsequent regulation and intermittent closures to foreign boats have helped the populations rebound in fits and starts. But by the turn of the century, more than fifty thousand tons were once again being hauled in annually, and the FAO finally declared the area "overfished." At that point, the regulators got serious, instituting an annual overall fishing quota of twenty-five thousand tons, various off-limits areas, and a prohibition against catching juveniles that weigh less than 450 grams (almost a pound). The rules seem to have worked, with much heftier animals being landed in recent years.

As Morocco's octopus stock took a dive in the early 1980s, before the stricter regulations took effect, Mauritania began upping its catches with more trawling. But the country's opportunism has also taken its toll. With spotty regulatory enforcement, their own octopus populations have been overfished as well. And it hasn't just been the locals. Plenty of European boats have been suspected of keeping underweight octopuses and fudging their catch numbers. Mauritania also signed an 86-million-euro agreement with the European Union in 2006 so that

European boats could fish for octopus in the Mauritanian waters. This extra trawl-based fishing pressure has raised concern about the area's being overexploited. Many saw the 25 percent drop in supply to Japan in the mid-2000s as a sign that the area is, indeed, being overfished. Seafood Watch suggests that the situation there is "critical" and discourages consumption of octopus from Mauritania.

Next in line seems to be Senegal, which is also getting into the octopus game. But it might have learned a lesson or two from other African countries and has already instituted size restrictions for catches and maintains an off season in an attempt to keep stocks healthy. Other places in Africa, including South Africa, are now also scaling up their octopus harvesting efforts.

With the ability to fish in North African waters, the Spanish fleet's octopus catch exploded, quadrupling from about five thousand tons a year in the 1950s to some twenty thousand tons in the 1990s. And after some rough years with broken agreements in the mid-2000s, Spain renewed agreements with Morocco and Mauritania and sealed a new one with Senegal.

European Union countries fishing in their own waters face relatively strict regulation. Since 2005, the European Union has mandated that octopuses be at least 450 grams (about a pound) in order to be kept, and some countries have much higher weight minimums. Jaime Otero helped to increase the local minimum size for creel catches in Galicia from at least 750 grams to 1 kilo, or from roughly 27 ounces to a little more than 2 pounds. This seemingly small change should allow many more female octopuses, which often reach sexual maturity at about 1 kilo, to avoid capture long enough to lay eggs and create a new generation of octopuses—to maintain future generations of fishermen. When I poked my head into the main Vigo fish market, I noticed a sign that reminded vendors of the 1 kilogram minimum weight. As many conservationists have pointed out, however, it's one thing to toss one of these *pequeños* back into the ocean after pulling it up in a creel; it's quite another when the too-small octopuses are caught in bottom-trawl nets (an ordeal they do not always come out of unharmed—or even alive).

Poster at the fish market reminding vendors that the minimum octopus size is one kilo.
(Katherine Harmon Courage)

And with high demand for small octopuses in the restaurant world (how was that octopus's garden cocktail?), many undersized ones do manage to squeeze through the regulatory nets.

Guerra and his colleagues work closely with local fishermen and the government to understand how local octopus populations can be managed better. And increasingly, fishermen do understand that it is important "to manage the sea," as Guerra says. One way to do that is to slowly institute restrictions on fishing areas that are important for octopus hatcheries, such as nearby Islas Cíes, which is now a national park. But it can be a battle, Guerra says, because like everyone else, the fishermen "want to get money. They want to be wealthy, as soon as possible."

Another option for local fishermen is to shuttle tourists instead of seafood. "They can get a lot of money doing this instead of fishing," Guerra says. And toting around sightseers could make for a much easier day than hauling up unwitting animals in all kinds of weather. "It's a very hard task to go to the sea," Guerra says. "They do it all year except the closed season," which is but a handful of weeks in the spring. With such small, flexible boats and lean crews, the fishermen can also switch from cephalopods to crabs to bivalves to fish during different parts of the years, changing gear—creel, long line, trammel net—as need be. But not every proud Galician fisherman is willing to turn in his creels for a boatload of Canadians.

A new organization has sprung up to help fishermen through tough times by turning their trade into a tourist experience in itself. Called Pescanatur, it allows paying sightseers/adventure seekers to ride along with fishermen as they take in their haul. The program is too new to tell if this hybrid business model might help fishermen stay afloat into the future—or just how many visitors are looking for the *full* frontline experience, like the one I received.

Pulpo Caldera

By the time darkness falls on Vigo again, I have regained my appetite and am ready for an octopus dinner at Don Quixote, a grand restaurant perched above the waterfront in the old part of the city. At the bottom end of the terrace, a giant copper caldron is steaming into the brisk evening breeze.

My waiter, Oscar, a clean-shaven young man with a good-old-boy 1950s haircut, explains that during the busiest parts of the summer, the restaurant will go through maybe a hundred pounds of octopus a day. They boil the day's octopus in the copper pot, as he says, "in the morning"—Spanish time (something like 12:30 or 1:00 p.m.). As soon as an octopus is tender, they take it out and place it on a covered platter that sits on a wooden plank above the caldron. The secret to octopus cooking, he says, is in the pot. If you boil it in a steel pot at home, it's likely to take thirty to forty minutes to become tender. But with a copper pot, it should take much less, he reports.*

I order my *pulpo* and watch an old waiter in a black bow tie shuffle over to the big caldron. He picks up a small wooden dish and dips it

* This Spanish suggestion seems akin to the Italian insistence that one should always put a wine cork in the water with a boiling octopus to tenderize it.

into the steaming water a few times. Then, from the platter, he selects an octopus, which he begins trimming into small bite-sized pieces with gleaming metal scissors. When the little wooden plate is covered high with these morsels, he dips a pitcher into the steaming water and slowly pours the water over the plate, warming up the octopus. Now, a few shakes of salt, a dash of paprika, and a generous drizzle of olive oil. And it's ready to serve.

Oscar volunteers that Vigo is okay for octopus, but the real place for octopus is the village of Bueu, where he happens to be from. "You know Bueu?" he asks. Oh yes, I know Bueu—where my ill-fated morning fishing trip had started. His favorite local method of preparing octopus is *pulpo caldera,* which is what they make on the Island of Ons—what I might have eaten with the fishermen, if I could have stomached lunch.

Pulpo Caldera (with *Ajada* Sauce)

Courtesy of Oscar, at Don Quixote

Boil an octopus with a whole peeled onion for about 30 minutes, or until tender enough to poke through with a fork.

Remove the octopus and add potatoes to boil until tender.

While the potatoes are boiling, sauté 6 to 10 cloves of garlic in olive oil, just to the point of becoming golden.

Remove the sautéed mixture from the flame and add paprika (a 50/50 sweet and hot blend is best) and then a splash of vinegar.

Carefully spoon out the garlic pieces and discard.

Remove potatoes and place on a dish with octopus.

Mix with a little water from the pot and spoon oil sauce over octopus and potatoes.

Each small place has its own take on the simple octopus dish. The next day, Jaime Otero and I drive to see more of the coastal fishing villages. The countryside is covered in small vineyards and little houses

Restaurant mural in Galicia advertising octopus.
(Katherine Harmon Courage)

with sepulcher-like granaries. Distant hills are topped with wind turbines, which catch the strong Atlantic gusts.

We visit A Illa de Arousa, a small island that was only recently connected to the mainland by a bridge. That island, of course, has its own boiled octopus recipe. To sample it, Otero takes me to a favorite local restaurant called Bar Clube not far from his family's seaside summer house. "*Expecialidad en pulpo*," the menu reads. Otero orders up another feast. We start with a delicious local Spanish pepper, fried and salted. Next comes cuttlefish in its own ink (which is remarkably different in texture from octopus—it's soft and cleaves under the fork more readily, almost like fish; the ink has a mineral, earthy essence to it). Finally, *pulpo estilo Illa* (the local version of *pulpo à la Gallega*), prepared in a special (though indiscernible to me) manner of this town.

As we drive on, we pass a statue of an octopus fisherman and his wife that greets motorists. "It tells you how deep inside the culture the octopus is," Otero says. We soon see a restaurant called Bar López, where a freshly painted mural on the side of the wall advertises its offerings with a big, lifelike floating octopus: "Put an octopus in your life," Otero paraphrases. Indeed, now it seems like we can hardly avoid it.

CHAPTER TWO

Tough and Tasty

The octopus is not only difficult to catch, it is also difficult to eat. Through the ages, cultures have developed seemingly bizarre methods for turning the tough meat into something that is, if not tender, at least supple enough to attempt mastication. Some people hang their octopuses out on lines in the sun; others bludgeon them on—or with—rocks. And at least one modern octopus distributor has mechanized the process with industrial octopus tumblers.

Nevertheless, the tough texture isn't the only obstacle people face. Octopus is not a naturally appealing dish for everyone. Many people (my husband included) simply cannot get past the appearance of octopus arms on the plate, suckers and all. When served up at home or on a platter at a restaurant, an octopus is at least technically tamed, but it still seems entirely alien—the suckers, the arms, disembodied, like a creature wrangled from another planet, another galaxy. Perhaps an aversion to octopus is just a prudish Puritan sensibility. People from much of the world over and even many of the scientists I spoke with seem to have no trouble facing an octopus on their plate. For so many of the other animals we eat, we go to great lengths to disassociate the dish from the live animal: pigs become pork, ham, or bacon (only occasionally are they served in whole form); cows are beef; and even deer become venison. Sea creatures, for some reason, seem not to undergo this sort of transformation—except where foreign names are adopted (calamari) or fish are rebranded to make them sound more appealing (Patagonian toothfish becomes Chilean sea bass, for example). Is it simply that there is often less animal to work with, to cut away into different shapes (into nuggets or chops)? Or do we lose a sense of discomfort because the thing that we are eating is so utterly dissimilar to ourselves?

Those of us who are initially tentative about cephalopod consumption might never understand what compelled people, throughout thousands of years of human history, to pull an octopus out of the water and decide that this odd floppy creature looked like a decent dinner. But it stands to reason why these animals have stayed on the menu. A 3-ounce serving of octopus meat has just 139 calories and 2 grams of fat. It's chockful of protein (25 grams), iron (45 percent of your daily recommended allowance), vitamin B_{12} (510 percent), and even copper (about 19 percent). Compare that to a roasted chicken breast, which has about the same number of calories and protein but has 3 grams of fat, just 6 percent of daily iron and B_{12} recommendations, and only about 3 percent of copper. Octopus totally trumps any chicken—and almost even chicken of the sea.

But before making a nutritious octopus dinner delicious, it's usually necessary to do a bit of battle with the octopus's anatomy. And that usually starts with a good beating.

Taming a Tough Supper

Octopuses are tough—and not just in the sense that they can take out sharks (both real and computer generated, as in *Mega Shark versus Giant Octopus*). They're almost pure muscle. With tridirectional muscles in the arms, they're a tad less supple than a well-marbled sirloin, to say the least (though certainly a lot more healthful). So over the centuries, people have been finding ways to make them a little easier on the jaw.

The classic tactic is beating the bejesus out of them on rocks. Many a tourist has been disturbed out of their beach-bathing stupor by the sight of a local smacking a dead octopus on the shore. The beachside brutality is a simple substitution for a meat-tenderizing mallet in the kitchen. It's so well accepted that one of my waiters in Vigo says that after catching one you must "hit it with a stone, thirty to forty times." He mentions this as casually as if suggesting one ought to fillet a fish before serving it.

For the squeamish midcentury-American palate, octopuses receive the splay, sauté, suck-it-up-and-swallow-it treatment so charmingly espoused by Irma Rombauer and Marion Rombauer Becker in the *Joy of Cooking*. To introduce the intrepid home chef to octopus and squid, they begin: "Both of these inkfish belong emphatically to the large category of horrendous-looking sea creatures that must be eaten to be appreciated."

They continue the inspiring preparation instructions:

> [Those] that are longer than 8 inches after cleaning need tenderizing. Pound them mercilessly on a solid surface. To prepare fresh octopus for cooking, make sure first that your victim is dead—by striking it a conclusive blow on the head.

So lovingly handled thus far, the octopus is then chopped, cooked, and dropped like yesterday's hash into what else but a mushroom-and-onion casserole. To be served with, they recommend—and I'm not kidding—creamed spinach. As offensive as this might sound to today's culinary purist, I admit I had some wonderful potato-and-octopus salad in La Spezia, Italy, that would have been right at home at an Ohio Lutheran church picnic—save for some errant suckers sticking out here and there.

With the wider spread of frozen food distribution, prepping an octopus for dinner has gotten a lot gentler. Octopuses are some of the few animals for which freezing does their texture a favor. The process of freezing breaks down cells in the tissues, making them much more tender. Most commercially caught octopus is now frozen as a matter of course, which makes distribution and consumption easier on everyone.

One adventurous entrepreneur in Brooklyn has taken tenderization to a new level, borrowing lessons from his ancestors back in southern Italy. Vincent "Vinny" Cutrone runs Octopus Garden, one of New York City's best-known octopus distributors. Cutrone is a small, jolly man with a high voice and glasses. His facility is in the Gravesend neighborhood of Brooklyn, amid old-school pizza joints and Italian bakeries.

There he chats up—mostly in Italian—customers who drive in from New Jersey and distributors who call in from Florida. His octopus and cuttlefish end up in the kitchens of some of the toniest restaurants in town.

When I arrive midmorning the day after New Year's Day, he has already closed up the retail side of the operation. But he's still kicking back in his office with a local Brooklyn chef, Bruno Milone, and an assistant. The three of them are at ease, shooting the breeze in Italian.

Cutrone graciously welcomes me in. My Italian from college is far too rusty to use, so they kindly switch to English. Milone, who wears a black Adidas tracksuit and white cooking clogs, says something about grappa and octopus. "He lost me," Cutrone says. "I can understand him better in Italian." Milone has heavy-lidded eyes and a few days' uneven stubble. "Nobody understands the octopus," he says in his heavily accented English. English is actually Milone's third language, after Italian and Spanish. He has been cooking for more than twenty-four years, having started when he was a teenager. Now he has risen through the ranks of bars, hotels, and restaurants to be executive chef at Marco Polo, a traditional Italian restaurant in Brooklyn's lovely Carroll Gardens neighborhood. He even travels back to Italy from time to time to teach at a culinary school.

Not ten minutes after I arrive, Cutrone is holding a raw baby octopus out for me to examine—its pale legs tucked over its body. The octopus is then whisked away to the office's kitchenette for a two-minute spin in the microwave. And then it's back. Cutrone hands over the piping-hot octopus on a paper party plate along with some plastic cutlery. I hesitate for a moment, still a little stunned, especially since it's only a little after 11 a.m. The cooked octopus is sitting in a pool of pale red liquid, which I reassure myself isn't actually blood, just excess that has cooked out and taken some of the skin pigment along with it. I am suddenly thankful that I shoveled down some granola before heading over, because microwaved octopus does not sound like the wisest first meal of the day—especially given my experience in Croatia. Cutrone explains, perhaps noticing my pause, that microwaved octopus is a normal snack around the office (along with microwaved cuttlefish, which we try later). I take a tentative bite. It's actually delicious. Tender, salty, and simple. A perfect, lean midday snack.

Cutrone next leads us into the main back room, where three big, round, horizontal metal drums, hooked up to external motors line one wall. On the other side of the room, a huge metal prep table is covered with bags of frozen octopus. Milone continues to talk about his latest octopus creations. Both he and Cutrone are boisterous natural salesmen—Milone giving the hard pitch and Cutrone the soft sell. As a result, the two frequently talk in simultaneous, overlapping, but-not-quite-congruous conversations.

In Puglia, where both their families are from, local fishermen who dive or spearfish for octopus bite the head to kill them, Cutrone informs me. Then they beat them in the surf, wash them, and serve them—often raw—right on the beach in fresh fish shacks.

In Brooklyn, things go a little differently.

"I'm continuing the traditions of the Mediterranean, from the area where we come from," Cutrone says, amid the gleaming machinery in the chill of the prep room. He's never fished for octopus himself, but that seems okay. He now has a massive freezer full of octopuses that have come to *him*.

"We're moving a lot of octopus," Cutrone says. Tons of it, actually. Each year they receive some 250,000 pounds of octopus, he notes. "It sounds like a lot, but we're actually doing that" much volume, Cutrone says. Most of it arrives cleaned, but they check it anyway and remove any unpleasant contents, such as remnants of the digestive tract or grit inside the mantles.

Workers come in at 5 a.m. to start the tenderizers churning. The drums have paddles attached inside, like a clothes dryer. They take about an hour to tenderize the octopus, and the water is changed a couple of times during the process. Workers add octopuses, ice, and salted water. Freshwater would change the quality of the meat, Cutrone notes. The water stays at about 3 percent salinity—roughly the same concentration as the ocean, Cutrone points out, so "the octopus is in its natural environment," more or less. Each drum can process five hundred pounds a day, so he can tumble as many as fifteen hundred pounds of octopus each day, five to six days a week. He says he's currently waiting for one New Jersey restaurateur to pick up five hundred pounds of tenderized octopus. "Some people, they can't live without it," he says. Around the shop, they say the octopus is perfectly okay to eat raw, but they're not allowed to market it that way. ("So we leave it up to the consumer to do whatever they want," Cutrone says, diplomatically.)

Standing around a metal prep table, we sample some of Milone's homemade octopus cocktail, which he proudly serves from a canning jar. It has delicate octopus arms, orange peel, red onion slices, raisins, white wine, and olive oil. It's tasty and a bit sweet. Cutrone and Milone note that it's perfect for a snack with a drink. But, alas, it's still before noon, so we settle for water—to be followed later by a shot of olive oil.

Most of Cutrone's octopuses, he says, come from the North Atlantic (mainly from, or at least by way of, Vigo, Spain) and the Canary Islands. Much of the Mediterranean fishing for export has stopped. "They're basically relying on Spain—Spain has this burden now," Cutrone says. "And the number of catches, the level of catches, are not there like in the past." Cutrone knows that much of the octopus is being fished from North Africa—Nigeria, Morocco, Mauritania, Libya, Senegal.

Cutrone bought into the octopus business in the late 1990s, and the name Octopus Garden "just clicked," he says. Before getting into octopus, he was, as he says, "diversified" (a portfolio that included music, furs, and real estate). Now his diversification extends mostly to cuttlefish and, occasionally, some squid. But demand for octopus meat has sustained him, continuing to rise over the past fifteen years. Octopuses from Cutrone's shop reach at least two hundred restaurants—some as far away as the West Coast and Bermuda.

Cutrone's operation handles all sizes of octopus, from a few ounces up to seven or eight pounds. (When I point out that the smaller ones seem to be restricted for fishing in many places, he says, "Well, whatever we're able to get here is not restricted by any American laws.") The best bet to buy for dinner, he says, is something around two pounds, which can be perfect in a salad. Or for a larger one, you can serve it *affogato* ("drowned") in a tomato sauce made with tomatoes, onions, and pepper. His favorite way to eat an octopus (aside from the quick and trusty microwave method) is just from the grill—perhaps parboiled then cooked over the flame with a little olive oil and lemon. He recommends boiling for the big ones, which tend to be a little tougher. The smallest octopuses, he says, can even be lightly breaded and fried, like calamari. Milone's specialty is octopus carpaccio, a traditional dish from around Milan.

Cutrone then leads us to the closed storefront area, where there are still several octopuses sitting out on ice. They're folded over, mouth side up, with their arms bent back, obscuring their mantles. Arranged as such, they look like strange sea stars or fleshy alien flowers. He shows off a few of the larger ones. And Milone talks him into cutting off an

Cutrone holds up freshly tenderized octopuses. (Katherine Harmon Courage)

arm of a big one to pop into the microwave to taste. Cutrone uses it for a brief biogastro lesson, pointing out the white tenderized meat in the middle and the mushy collagen holding the skin on the outside. Never accept an octopus without its skin, he says. "Send it back!"

The nuked arm returns, red and vibrant on a Styrofoam dish. I take a bite, which is a bit chewy, and I get coaxed into chasing it with a shot of olive oil. It's still before noon and we've had one small microwaved octopus, octopus cocktail out of a jar, and an octopus arm (in addition to microwaved cuttlefish, *baccalà*, and a tuna belly spread of Bruno's fixing). Does Cutrone himself ever get tired of octopus? "No," he says, emphatically. "It's something very healthy to eat." To these fellas, the octopus is not a marvel of neural complexity or behavioral wizardry. It is what it is to most of us and other larger animals: mostly just a nutritious, delicious meal.

To conclude the little tour, Cutrone shows me a short film called *Amor Pulpo,* which won Best Food Porn Film at the 2011 Chicago Food Film Festival. It features an elegant temptress slowly preparing for a sexy date. Meanwhile, an unfortunate octopus is being caught, tenderized, prepared, and grilled. Finally, the two meet at a restaurant for an intimate dinner. That octopus, Cutrone says, proudly pointing out the tenderizing scene, was one of his.

Before I can leave, he insists that I take some frozen octopus with me. And so I wind up with a package of *moscardini polipetti* (tenderized baby octopus) in my bag to take back with me on the subway, making for a close second (after a live lobster—it's a long story) in a list of strange things I've trekked home in my purse on the train. "We're not rushing you," he says, as I head for the door. "As you see, we're just talking octopus here."

Meaty Cephalopod Mysteries

As an octopus consumer—unless you're a true epicurean—it's tough enough to know what type of octopus, exactly, you're chewing on. When you go to the grocery store or fish market, an octopus marked "common octopus" or "*Octopus vulgaris,*" says Nikolaos Schizas, "could be anything." Schizas works at the Isla Magueyes Laboratory of the University of Puerto Rico, on a small island just off the coast of La Parguera. I meet him late one afternoon in March in his semichaotic office, where he is sitting in front of two computer monitors and drinking from a stained coffee mug decorated with Christmas holly.

He and a graduate student have been studying the genetic differences of various common octopus populations around the globe. And he proposes turning some of that high-tech sequencing to the local fish market. DNA analysis has turned up plenty of falsely advertised fish species (tuna is a common sushi restaurant sleight-of-species offering, but the problem extends all the way up to faux shark and down to changeling catfish). With just a few hundred dollars, and a few errands, he's pretty sure he could get a quick bead on what is actually being sold as "common octopus," he says. "I think the results would be all over the place," he says in his native Greek accent with a jovial laugh.

And sometimes you might not even be getting octopus at all. Later that evening, Roy Armstrong, the director of the University of Puerto Rico marine field station's Bio-optical Oceanography Laboratory and my unofficial guide in La Parguera, was kind enough to throw an illustrative dinner party in honor of my visit. He and his wife invited some of their friends from the university—a mix of born-and-bread Puerto Ricans and stateside transplants. Armstrong volunteered to showcase a recent affront he had noticed in the frozen foods section of his grocery store.

In Puerto Rico, octopus is a common enough ingredient that you can buy packages of precooked frozen chopped-up arms—like frozen shrimp. He had found, however, that some of the frozen stuff he was getting was, well, terrible. That's when he took a closer look at the bag. It had a cute cartoon drawing of an octopus on the top, but instead of saying *"pulpo,"* it read *"pupo"*—in other words, not *exactly* octopus. It would be like labeling the bag "octopos" or "octupus" or "octoplus" or anything not "octopus" (kind of like my dad's favorite knockoff sunglasses in the 1990s: a pair of "Okeys," which was especially fitting for an Oklahoman).

But if it isn't *pulpo*, just what is in this mystery bag? According to the fine print on the bag, it does contain "cephalopod," but without a fisheries background you might be at a loss. Fortunately, I am in good, knowledgeable, and outspoken company. The group informs me that it's actually squid, caught off the coast of Chile, where these ten-appendaged cephalopods are prevalent and cheap.

So what? you ask. One cephalopod arm should be as good as the next, and hey, calamari can be pretty darn delicious. Well, Armstrong, ever the scientist, has devised an experiment. He makes—in identical methods—boiled frozen *pulpo* (from a bag that did say "frozen octopus" and whose contents came from Indonesia) and boiled frozen *pupo*. He prepares both *a feira*, with olive oil, salt, and paprika, just like in Spain. Served right next to each other, one has a deeper red color, the other is a lighter shade throughout its meat.

Armstrong offers his real *pulpo* dish.
(Katherine Harmon Courage)

I approach the two plates of cephalopod. On the right are chunks of familiar-looking, red-skinned arms, lined with suckers. On the left are paler, shorter, thicker chunks of arm, with only small patches of dark skin. I taste the darker one first. It's a little chewy, but relatively tender and indeed reminiscent of some of the octopus I had had in Vigo. Then for a bite of the lighter one. *Blech*! It is tough and fishy and weird. Not octopus at all.*

Although some scientists, experts, and octopus admirers refuse to touch an octopus dish, plenty of people who revere them do also eat them. In a 1935 essay called "Feet in the Water," filmmaker and biologist Jean Painlevé was complaining about the many technical difficulties of trying to film his underwater subjects. "But there are consolations: the greatest being the ability to eat one's actors," he wrote. "Of course there is much to consider before tossing them in the pan," he continued. "There are one hundred seven varieties of bouillabaisse to choose from. Should one use garlic or not? Prepare it au gratin? Sautee in red wine? Add sardines? Classical gourmets may be offended, but the bouillabaisse of Marseilles cannot be imitated, so anything is allowed." That solves that problem. Just go for it.

Prepared with Love

Gythio is the type of town where if you round a corner too quickly, you're likely to end up getting a face full of wet octopus arms. Just about every fourth restaurant in this Peloponnesian fishing village has a line strung in front, heavy with dangling octopuses drying in the strong Mediterranean sun like the morning wash. The practice is not some morbid advertisement but rather a culinary strategy; the drying helps get them ready to throw on the grill for dinner.

* Not all squid is so foul, of course. And even scientific samples can be tasty, according to some truly won-over researchers. Lydia Mäthger, in Woods Hole, describes a previous lab where she worked on squid. "The squid were always shared between the different labs for the different research projects," she recalls. After all the useful parts—eyes, brain, skin—had been divvied up among the researchers, any extra muscle would get put in the freezer. Then, once in a while, one of her colleagues from Portugal would cook up the research discards into a squid curry. "It was excellent," Mäthger says. Years later, she still ranks it at the top of her list for cephalopod dishes.

Octopuses hang to dry outside a restaurant in Gythio. (Katherine Harmon Courage)

Gythio (also called Gythion, Gytheio, or Ytheio) is billed by a smattering of travel books and Web sites as "the octopus capital of Greece" (not that anyone had bothered, apparently, to inform the residents of this distinction). For a capital of anything, however, it is not terribly easy to reach. It requires catching a flight to Athens, which according to Madrid airport monitors does not exist, then an Athens city bus to the unlovely long-distance bus station, and finally—not the last Gythio bus, because that left an hour before—a late-night bus to Sparta, which will deliver you to a strip of utilitarian hotels where you can spend the night and catch a bus to Gythio the following morning.

Once you arrive in Gythio, the little town trickles down along the sea, tucking into one of the two big southern bays of the Peloponnese on the eastern edge of the Mani peninsula. It's wrapped in by an ancient Roman amphitheater on the north and, on the south, the peaceful islet of Cranae. It was on that small island, now connected to the mainland with a short, unromantic causeway, that Paris and Helen supposedly consummated the affair that would lead to the Trojan War. Cranae, or Kranai (also known as Marathonisi), like just about everything in Greece, has at least two or three entirely different names—everything, that is, except octopus, which is, simply, thankfully, just *oktapodi.*

During my first day in town I quickly become aware that there is no English to be found—anywhere. So there go my hopes of long, thoughtful conversations with local chefs and residents about their relationship with octopus. With my dozen words of Greek (*hello, thank you, octopus, bus station, bye*), there would be but simple verbal exchanges. So I resign myself to a more gustatory inquisition.

The proprietors of the harborside restaurants seem perplexed—beyond the language barrier—when I ask them where their octopus is from. "Here," most of them eventually answer with a funny look on their faces, as if I had just asked them if they were proprietors of restaurants in a small Greek fishing town. When I ask one owner *when* their octopus was caught, she responds, after some thinking: "Tomorrow—

Freshly caught octopus arms hang to dry in Gythio's harbor.
(Katherine Harmon Courage)

today and a day." Although perhaps she means "yesterday," I take it to mean the grilled octopus on my plate is *so* fresh it might well have been caught in the future.

I discover that my innkeeper, Odile, is from Brittany. So between my flagging French and her fair English, we can actually kind of communicate. She sets me up to speak with the chef at the attached restaurant who loves to cook "the octopus," she says. Her daughter will serve as a halfway English/French translator. On the appointed day, my young translator and I walk across the busy main street separating the restaurant from rows of seaside tables. There sits a weathered Greek man smoking Winstons with an empty red espresso cup in front of him. This is the octopus chef, Takis Zaloumis.

The three of us manage a slow, halting conversation with Greek, English, and a little French, punctuated by long pauses to let the noisy motorcycles and trucks rumble by on the road. Zaloumis has thick, strong hands. His deeply creased face has merry blue eyes that are mostly hidden behind a perpetual squint against the strong southern Peloponnesian sun. He has been cooking for thirty years, he says. He seems slightly amused that this American writer is asking such embarrassingly obvious questions about octopus. But having finished his coffee, he sits back to enjoy his cigarette and seems content to indulge me.

He explains that the first important step in cooking octopus happens before it even reaches the market. In Greek to English, the brutal process of traditional tenderizing is described in gentle words: "The fisherman, he takes the octopus and puts it down—to make it softer" (that is, he beats the hell out of it).

Zaloumis gets his octopus mainly from a fisherman named Takis Anagnostakos, who also happens to be his neighbor. Only when the fisherman Takis is without octopus does the chef Takis go to the market to get the goods (these store-bought octopuses also have the benefit of being pretenderized, he says). Then they hang the arms in the sun to dry out (for maybe a day or so), which makes them cook better on the grill, I'm told.

Diplomatic Zaloumis won't profess a preference for a particular octopus dish. "He likes them all," Odile's daughter translates. "The grill, boiled, or with spaghetti—it's different, the taste," she says. He offers his simple recipe for octopus salad, the Greek appetizer mainstay:

Octopus Salad

Courtesy of Takis Zaloumis, chef at Saga Restaurant, Gythio

Clean the octopus.

Put it in water with a little vinegar and boil for about an hour.

When it's done, put the octopus in a bowl, add oil, vinegar, and a little bit of the water.

But Zaloumis does not divulge the recipe for his more indulgent octopus dish, the one that he will make for me that evening. "In the night," our translator says, "he will cook for you the octopus the way he wants you to eat it. What time do you want to come?" Zaloumis suggests seven o'clock. It's a date.

That evening, I arrive for my date with the octopus. I'm seated at a table, right along the edge of the sea, where I can hear the water sloshing and see the urchins attached below. The sun is just starting to set in the glowing sky, with Cranae as a backdrop. And I am ready for my octopus feast. First comes an astoundingly fresh Greek salad—unlike anything I've ever had before, with fresh chunks of flavorful tomato, feta, and herbs. Then the main attraction—an exquisite dish of glazed octopus arms, white rice, and sautéed red peppers, topped with fresh baby basil leaves and mint. The octopus is tender and perfect. The arms taper down to the most delicate tips, which curl into a soft spiral. When you pick one up, it unfurls slowly, like a secret. After a few bites, I look up across the street and see Zaloumis sitting at a side table taking a break, checking on me from a distance. I wave and give him a lame thumbs-up. He smiles and gives me a thumbs-up back.

Zaloumis's octopus masterpiece. (Katherine Harmon Courage)

That the animal can so beguile or disgust diners and so vex or vindicate chefs is perhaps part of the octopus's appeal. While it is a meaty staple in many places in the world, it is still a novelty in others. But octopuses are even more bizarre than the most creative culinary presentations or outlandish tenderization strategies would let on. Their oddities in life go way beyond any weirdness they might display in death—or dinner.

CHAPTER THREE

A Strange Animal

For all of the thousands of years we have been catching and eating octopuses, the animals themselves have remained strange to us. They have a lot more to offer than high seas fishing adventures and elegant dinners. They live in another world and have bodies to match. With hundreds of different species, some can even surprise professed octopus experts with their totally mind-bending weirdness. To start to understand the octopus intimately, one must look past its repulsive—or delicious—qualities. One must use a naturalist's eye to find its place in a sea of outlandish relatives, examine its odd inner workings, and get to know its curious habits.

First, however, we must avoid being sucked into the clutches of those many legendary cephalopods that are *not* octopuses. For example, it was not an octopus that attacked Captain Nemo's *Nautilus*. Nor was it one that assaulted a ship in *Moby-Dick* or wrestled James Bond in *Dr. No*. No, these attacks were the work of—albeit fictional—squid.

Both of these "army" creatures—octopus and squid—are often intertwined in the popular imagination, not to mention restaurant logos and children's book illustrations. And they're united in the mythological kraken, that monstrous and malevolent squid-octopus beast that supposedly lived in deep waters and sought out ships and sailors as snacks. Such fearsome fame has elevated the actual cephalopods to animals of almost unnatural, unknowable mystique. Not a single actual octopus, however, has been documented to be large enough to sink a proper ship—or anything much larger than a canoe.

Once one gets past the primordial fear of submerged entanglement in so many arms, one learns that the two animals are quite different.

Pierre Denys de Montfort's "Poulpe Colossal" attacks a merchant ship.
(Pierre Denys de Montfort, 1810)

Squid and octopus both belong to the cephalopod class. But the two orders—teuthida and octopoda—split ages ago, possibly sometime deep in the Paleozoic era, when the two-tentacled squid was relegated to a ten-extremity fate and the octopus, to a refined eight-armed existence. The squid's life is quite different from the octopus's. Squid jet around—often in schools—in the open seas, while most octopuses usually prefer to crawl—mostly alone—along the bottom (although some elusive species live in the open ocean). Some squid use their scary, hook-covered tentacles to lash out and capture prey, while octopuses rely on their dexterous suckers and strong arms to grab dinner.

Dangling around octopus and squid from the cephalopod family tree are a couple of less famous cousins. Shelled nautiluses drift around the seas in much the same form as their ancestors that were fossilized some 500 million years ago. They are the only cephalopod to have retained a protective shell, which they use for shelter and buoyancy. They've outdone themselves on the limb front, boasting ninety or so whiskery tentacles around their hidden mouths. Also on the family tree are the single-boned cuttlefish. They aren't actually all that cuddly, but they have been shown to be quite clever—and can sometimes even outcamouflage an octopus. All in all, the few extant cephalopods make for a cozy little class.

A Diverse Order

The scope of the different octopus species alone is enough to boggle the human mind. Researchers have started taking advantage of genetic tools to get a deeper sense of the octopod group's layout. But even with DNA sequencing technology in hand, we have not yet been able to find and describe each species and subspecies. Eric Hochberg, who recently retired as curator of invertebrate zoology at the Santa Barbara Museum of Natural History in California, suggests that even though three hundred or so octopus species have been named, "probably there are at least that many—or more—that are undescribed." To find all of the world's octopod species one needs not only to scour the seafloor near and far but also to survey the vast midwater world between the surface and bottom. And so far, such an exhaustive search has proved to be too expensive and time-consuming to undertake. But for now, we can take a brief tour of the more remarkable species that *are* known to science.

The little argonaut (*Argonauta*) is one of the mysterious midwater octopods that lives out its life suspended in the water column. It appears more nautilus-like than most other shell-less cephalopods. The mature female argonaut creates a thin egg case around itself and uses pockets of air captured in the shell to hover in midwater. This genus captivated Aristotle, with its shell that he likened to a sail in the breeze, and its arms "as rudder-oars." But as in Aristotle's time, when "knowledge from observation [was] not yet satisfactory," as he noted, we still know little about this octopod and how it evolved to construct these strange structures.

Much more well studied are shallow-water bottom dwellers such as the California mudflat octopus *(Octopus bimaculoides)*. Found in the wild on the West Coast, it's also a common resident in research labs and a popular breed of pet octopus. It can live for a couple of years and can grow to be a couple of feet long from mantle to arm tip.

The North Pacific flapjack devilfish (*Opisthoteuthis californiana*) should be familiar to millions from the Pearl character in Disney's *Finding Nemo.* It is one of the flattest known species in the entire class of cephalopods. In real life, it barely looks like an octopus at all, with fins, a blobby web, and short stumpy arms.

The giant Pacific octopus (*Enteroctopus dofleini*), on the other hand, better lives up to the iconic octopus image. Its besuckered arms can reach a span of more than 20 feet, and it can weigh upward of 150 pounds. Because of its size—and its relatively high tolerance for life in captivity—it is the species commonly on view at public aquariums. For that reason, it is also the most prone to making headlines by opening tank valves, disassembling expensive equipment, and generally wreaking havoc in labs and aquariums. Although many of us—especially those of us who prefer not to spend more time than necessary submerged in bodies of water much larger or darker than a backyard swimming pool—would probably prefer not to meet a giant Pacific octopus in the open ocean, it is hardly the most dangerous.

That honor goes to the tiny blue-ringed or blue-lined octopus genus (*Hapalochlaena*). These small cephalopods are not just an invention of Ian Fleming. The malignant muse of *Octopussy* (the 1966 James Bond short story by Fleming and the 1983 film with Roger Moore) is a real—if physically unassuming—menace. At just one to two inches in total length, it might look harmless and even fanciful crawling around in a tide pool. But each one packs a poisonous punch with venom that can kill an adult human within minutes. There are numerous anecdotes about people dying after a bite from this octopus's tiny beak. And many more mysterious seaside deaths in Australia and the Indo-Pacific have likely been caused by these cute creatures. Humans, of course, are not their main target. They use their potent neurotoxin to stun prey and, occasionally, mates. In a truly femme-fatale move, one female blue-ringed octopus was observed killing (and even eating) a male after mating with him.

More recent research has revealed that just about all octopuses (and cuttlefish, as well as some species of squid) are at least a little venomous. Most are no threat to humans, but for potential prey they have the kiss of death. This protein-based poison can lead to paralysis and other nervous system malfunctions, leaving a crab or clam helpless in an octopus's clutches.

Telling the hundreds of octopus species apart can be difficult. Even the showy blue-ringed octopus comes in a few different species. And within a singular species, octopuses can look so different as to have fooled scientists into thinking they were only distant relations. The blanket octopus (*O. tremoctopus*), for example, has one of the most comically extreme examples of sexual dimorphism in all of the animal kingdom, and it left scientists searching for the male sex of this species for ages. The females have long, luxurious, fleshy webs connecting most of their front four arms. These gals can grow to be some six and a half feet long and weigh some twenty-two pounds. Males of this species, on the other hand, are not as well endowed. Their whole bodies are less than an inch long and less than a tenth of an ounce in weight. This odd pairing would be the equivalent of an average woman with a walnut-sized guy—or an average Joe with a gal six times bigger than towering Allison Hayes in the *Attack of the 50 Foot Woman*. Rather than risk a classic act of copulation, which could be awkward and probably a little dangerous, this minuscule male passes off a packet of his, er, materials to his large lady friend along with *one of his arms*. She stores his little gift in her mantle cavity for later, when she's ready to lay eggs.

The blanket octopus is not the only octopod to pair off so distantly. Most male octopuses, in fact, use a specialized arm (which happens to be their third, so go ahead and insert your third-leg joke here) to convey their genetic materials to the female. Some species insert this specialized arm into the female's mantle, but plenty of male octopods sever this makeshift member and give it all to the female, part and parcel. Losing an arm for the privilege of procreation might seem a bit extreme, but it's okay, because octopuses can regrow lost or damaged appendages. And most of them die soon after they procreate, but we'll get to that later.

Odd Anatomy

Octopuses can be difficult to get a handle on—intellectually and, well, literally. Their bodies are squishable and their arms are made of a nifty biological mechanism known as a muscular hydrostat. Thanks to these features, an octopus can change its dimensions dramatically while its total volume remains constant. Like the human tongue or a Stretch Armstrong toy, it can lengthen, contract, and contort. This allows them to squeeze through some seemingly impossibly tight spaces, hence the name of the octopus toy character in *Toy Story 3*: Stretch. A giant Pacific octopus, for example, can squeeze itself through anything smaller than its hard beak and braincase, which can mean slinking through an opening little larger than an inch in diameter.

Jean Painlevé likened this quality to pliant chewing gum. In fact, these bewitching animals were the reason he got into making science films for the public in the first place. It was "to convey my passion for the octopus," he told *Libération* in a 1986 interview. "It was a dream I'd had ever since meeting one during a childhood vacation in 1911." His first popular short, a ten-minute semisurrealist silent film in black and white, was *The Octopus* (1928), which he made for both education and entertainment. In this film we watch a small octopus slowly sliding out of a tiny hole in a handheld fishing net. And in one seaside studio mishap, his eight-armed star sneaked out of its tank, slithered under a door, and crawled out onto a high window ledge, from whence it continued its slinking. No report on how the octopus fared, but the sunbathers below were apparently rather surprised.

Inside all of that stretchiness, if you could examine them, you'd find that most octopuses share the same basic body plan: arms on the bottom, mantle on the top. The mantle is just a fancy name for the big, headlike body blob that rests on top of the arms. The "head"—the part that holds the eyes, central brain, and so on—is closest to the arms, not far from the mouth. The mouth is hidden away at the center of the eight arms. But be careful up there—it has a sharp beak and a scary toothed radula for drilling into hard shells. This chitin structure, awkwardly positioned on the octopus's underside, at the center of all of its appendages, almost evokes the strange myth of a *vagina dentata*. (But don't worry; because the female octopus accepts sperm from a male directly into her mantle, she keeps the chance for love bites to a minimum.)

Behind the head, in the mantle, are the digestive tract and other internal organs, including three hearts. (That's right. *Three hearts.*) The octopus breathes by sucking in water through gill slits, or lamellae, and exhaling it through the funnel, or siphon, which is the spout at the base of the body. This is the same tube it can move to jet around if it feels like swimming—or squirting an irritating scientist. It's also connected to the end of the digestive tract (that is, it also acts as a pooper).

Even back in the fourth century BCE, Aristotle already had the octopus's basic biology down pretty well, as he describes in his *History of Animals*:

> After or at the back of the mouth comes a long and narrow oesophagus, and close after that a crop or craw, large and spherical, like that of a bird; then comes the stomach, like the fourth stomach in ruminants; and the shape of it resembles the spiral convolution in the trumpet-shell; from the stomach there goes back again, in the direction of the mouth.

To get an even more intimate internal view of the octopus, one of the best places to start is, perhaps surprisingly, in the Midwest. In zoology curator Janet Voight's office at the Field Museum in Chicago, shelves and cabinets and countertops are crammed with jars of preserved octopus specimens. She and I meet on a hot August day at the museum's entrance perched above the decidedly octopus-less Lake Michigan. The hallways of the old institution smell like the friendly, musty basement library stacks from my undergraduate days. Voight's sprawling office is cluttered with the posters, books, and dissecting materials of an earnest naturalist.

Although it seems odd to be surrounded by so many octopuses in Illinois these days, the area wasn't always so inhospitable to the creatures. In fact, during the Carboniferous era, some 300 million years ago, a shallow sea covered much of the center of the country (and also left behind the fossilized marine shells I found near my childhood Oklahoma home). The oldest-known octopus specimen, *Pohlsepia*, resides at the Field Museum and was found right there in Illinois, just an hour and ten minutes' drive to the southwest, in the famous Mazon Creek fossil beds.

Voight has spent decades collecting more contemporary (though almost as hard to find) octopuses. To do this, she has had to go much farther afield—to some of the most hard-to-reach spots on earth, including the hydrothermal vents. As a result, her collection includes blanched, ghostly white specimens from deep seas.

She takes out a jar of baby *Graneledone* octopuses that were collected almost four thousand feet below the surface. They have pronounced white spots from a texture lingering beneath their skin. Another specimen she removes from a jar was her onetime pet someone had given her as a gift. His name was Bubba. I looked for any extra tenderness as she flipped back the precut mantle of the preserved octopus to display the internal organs. I see only the elegance of a practiced biologist's hands as she points out the small black pouch behind the rectum that is the ink sac, the big area that she calls "the analog of the liver," and the "terminal organ" of the male (its third arm).

Each of the octopus's eight appendages might look the same to us, especially if they're served up on a dinner plate. But octopuses seem to use some of their arms for different tasks. When trolling the seafloor (most octopuses' preferred mode of locomotion), octopuses will often use the back arms for walking and the front ones to feel around for food. Just check out Ursula in the 1989 Disney film *The Little Mermaid*, who, some contend, was modeled on old footage of octopuses filmed by ocean explorer Jacques Cousteau. In the lab, when offered an object, a real octopus often favors a left or right arm (usually of the first or second pair), thus shown to have a sort of handless "handedness" (or, perhaps more accurately, an armedness).

To keep those eight wonderful muscular arms going, the octopus has developed some amazing physiological solutions—both macro and molecular—over the eons.

The octopus might technically be cold-blooded and the basis for all sorts of fictional villains, but no one can ever accuse a real octopus of being heartless, as it comes with three.

One main heart does most of the work for the organs in the mantle, while two ancillary hearts move blood past the gills. With these backup tickers, the octopus can have a bit more flexibility in its beats. But that does not mean that its cardiovascular system is superpowered. The main heart, for instance, stops when the octopus is actively swimming. This means that it tires quickly and prefers to crawl along the seafloor for most of its errands rather than jet around.

Three hearts might seem like an extraordinary evolutionary development, and when you consider that plenty of the octopus's close relatives (such as clams) have only one heart and don't even have veins and arteries to carry the blood around their bodies in an orderly fashion, the octopus's circulatory system looks downright astounding.

Despite the octopus's generally lonely life, when it does sense love, the male's heart literally skips a beat. Zoologist Martin Wells, of Cambridge University, discovered this telltale behavior in the 1970s, when he introduced a female into a male octopus's tank. And it turned out that love—or lust—actually made the male's heart skip a *few* beats. And it's almost too coincidental that these missed beats happened at telling times: when the female arrived, when the two got closer, when they touched, and when the male passed over its sperm packets. And although our (singular) hearts tend to race with the introduction of a little romance, these offbeat hearts are a strangely touching parallel.

These multihearted creatures are even invoked in the name of a human heart condition. Called takotsubo cardiomyopathy (*takotsubo* being Japanese for "octopus trap"), it is known more commonly as broken-heart syndrome. In this temporary illness, the left portion of the heart enlarges and takes on an oblong shape, resembling the vaselike traditional Japanese octopus trap. Whether or not the octopus itself ever gets its heart broken we may never know—and no cephalopod-composed sonnets have ever been found.

The strangeness of the octopus's inner workings does not stop with its three hearts. Cut open an octopus and it bleeds blue.

Recall for a moment the brightly colored anatomical diagrams from middle school biology books or classroom posters that show the two sides of the human circulatory system: The red arteries carry well-oxygenated blood out to the body, and the blue veins carry the oxygen-depleted blood back to the heart and lungs to be replenished. Thus, as the logic goes, when you get a cut—even if it's on a vein, say one of the bluish lines on the back of your hand—as soon as the blood hits the oxygen-rich air, it turns bright red.

This easy color coding falls apart when octopuses enter the picture. When oxygenated, octopus blood runs blue, fading to a clearer hue when it lacks oxygen. Our blood and that of many other animals is red because it is full of iron. We depend on iron in our blood to create hemoglobin, the protein that helps to transport oxygen. At low oxygen levels, however, such as high altitudes, the effectiveness of this system starts to wane. Octopuses, along with some other dwellers of the deep such as horseshoe crabs and other mollusks, have solved the low-oxygen problem by developing a totally different blood base called hemocyanin, which is infused with copper rather than iron.

This hemocyanin and the "darn blue blood," as Roger Hanlon, a senior scientist at the Marine Biological Laboratory in Woods Hole and a well-known cephalopod expert, describes it, has left the octopus rather vulnerable to changes in water acidity. The Bohr shift—the rate of oxygen-carrying capacity over a span of pH levels—of the octopus is quite steep, meaning that just a small dip in pH leaves the hemocyanin much less capable of carrying oxygen in the blood. With a low enough pH the animal won't be able to get enough oxygen to its tissues and will essentially suffocate. So in comparison with animals that use hemoglobin, as fishes do, "the octopus and the cephalopods can't handle these shifts" very well at all, Hanlon notes. And that could turn out to be a pretty big limitation, especially in the face of climate change–induced ocean acidification.

The octopus circulatory system has apparently served it well so far in the seas, but as Hanlon notes, "one of the big failures of the cephalopods was that they never got on land," he says, citing Wells's work. Or, at least, they never made it onto land for more than brief lurching errands between tide pools.* They need full marine salt water in order for their kidneys to be able to conduct proper ion exchange.

Even inside an octopus's organs—just like in our own bodies—there is a whole ecosystem of life. As we have a massive microbiome of bacteria living in our guts, octopuses have a host of organisms living in theirs.

Eric Hochberg got into the octopus field via their kidneys—more specifically, the parasites that live in their kidneys. (Perhaps a larger claim to fame, however, is the time he spent in the Virgin Islands living underwater for three weeks in 1971. "I was really trying to understand what squid and octopus were there—and trying to just be able to be in a place on the bottom and follow these animals around," he explains. "It was sort of a jingly time.")

He noted in a 1983 paper that "almost without exception all large, mature cephalopods are infected with parasites. Viruses, bacteria, fungi, three phyla of protists, a phylum of mesozoans, and six phyla of metazoans have been recorded. Parasites have been recovered from almost all the tissues and organs of cephalopods," but they are most commonly found in their gills, muscles, digestive tract, and their kidneys. To that last location, as Hochberg pointed out in his earlier 1982 paper, "The 'Kidneys' of Cephalopods: A Unique Habitat for Parasites"—I'll say—"the kidneys of any organism are an unusual place to find parasites, and in fact in the entire animal kingdom this organ has rarely been exploited." This, he goes on to say, is a "rather remarkable situation."

But it was not altogether a new discovery. In fact, the first reference

* Or if the creatures of the online hoax Pacific Northwest tree octopus (*Octopus paxarbolis*) were to be believed, also occasionally in temperate rainforests.

to an octopus parasite dates back to at least the eighteenth century. In 1787, Italian naturalist Filippo Cavolini, found in octopus kidneys "little infusorial organisms, shaped like eels, having a muzzle with a trembling head, darting, dividing themselves into many positions."

Scientists have been trying to understand these strange organisms ever since. "We know what's going on inside the octopus kidney," Hochberg says in his excitable voice when he speaks with me on the phone from California. But we still don't know exactly how or when the organisms get there—whether it's when the octopus is a hatchling, juvenile, or young adult.

Octopus kidneys are not exactly like your kidneys or mine. They're a bit of a misnomer and are actually a more integral part of the digestive tract than ours are. And "probably 'parasite' is not the best word," Hochberg says of the beasties. The organisms don't really seem to be doing the octopus much harm. On the other hand, "from what we can tell, they don't seem to have a beneficial function," Hochberg says. (Many species of octopus, squid, and cuttlefish can also become infected with nematodes, which can be harmful.) So why, you ask, are octopus kidney parasites important?

From these nonparasite parasites, there are actually loads of lessons to be learned. By identifying an internal parasite, for example, you can also help to identify the species of the host animal. "That's another clue that we use for description," he says. And on a bigger scale, which perhaps might make only a true taxonomist marvel, squid species seem to have parasites similar to those of octopuses, but theirs are protozoans rather than mesozoans. And before you start flipping ahead to the next section, let me just point out that these are organisms from a completely different *phylum*, like, us-and-fungi different. But these different buggers have an almost identical life cycle, size, and shape, Hochberg points out. "So there! It's strongly convergent in their evolution to living in the kidneys and undergoing a life cycle in an animal—a short-lived animal," he says, almost in one breath. This seemingly small difference between octopus and squid parasites underscores their rich and different evolutionary histories *and* their stubbornly similar anatomy. That they could each play host to these organisms that, over the millennia, evolved to be so similar, is a reminder that no matter how anomalous these animals might seem to us, they come from long lines of predecessors whose anatomies fit—and created—crucial ecological niches.

Explosive Growth

These cool-blooded (poikilothermic) animals are incredibly efficient with their calories. Because they rely on the ambient water to regulate their body temperatures, they're not busy burning off their food to keep warm the way we are. That means they can put much of their consumed energy to use in bulking up. "They are amazingly good at converting protein from the food they eat into their own body mass," marine biologist James Wood says. The rest of the energy goes into respiration, activity, and getting ready to reproduce.

Many areas have their legendary records for biggest octopus. In the sleepy Greek village of Gythio, the chef Takis Zaloumis said he had heard of a local octopus being caught that weighed twenty-two pounds, although the biggest one he had ever seen was fifteen pounds.

The largest octopus to have been reliably recorded was a live giant Pacific octopus, which weighed in at about 156 pounds. One dead massive seven-armed octopus (*Haliphron atlanticus*, so called because the males' hectocotylus arm is usually hidden in a sac) was dragged up from the deep off the coast of New Zealand in 2002. Although it was a carcass and not entirely intact, it still weighed 134 pounds and stretched some 9.5 feet from mantle to arm tip.

But even the rumor of a giant Pacific octopus that reached 600 pounds (with an arm span of about 30 feet) doesn't hold a candle to the giant Jameson Irish Whiskey–loving octopus from the commercial. That fictional creature, in turn, would have trouble competing against the massive kraken in the 2000 movie *Octopus* (subtitle: *A Cruise to Hell*) that starts attacking ships and subs to feast on their human occupants, plunging them into its terrible toothy radula. And the real McCoys would most certainly never be big enough to take out *Octopus*, the too-big-to-be-sexy megayacht owned by Microsoft cofounder Paul Allen, which measures in at 414 feet and comes equipped with two helicopters and two submarines—one of which is even designed to do deep-ocean research.

A common octopus in the Inland Sea of Japan might weigh about thirty-five ounces at just four months of age and will commonly grow to about six and a half pounds by adulthood, a year or two later. Thanks to its healthy appetite and relatively long life span, the giant Pacific octopus, which starts out about the size of a rice grain when it hatches, often develops arm spans of more than twelve feet in its three to five years.

Lab experiments showed that Hawaiian octopus juveniles, for example, could gain more than 4 percent of their body weight each day. (The size record was set for Hawaiian octopuses in 2000, when one weighing some nineteen pounds was caught). For Voight's dissertation, she found one species of octopus (Diguet's pygmy octopus, *Octopus digueti*) that grew from one four hundredth of a gram (about fifteen hundredths of an ounce) to forty grams (about one and a half ounces) in six months. That's a thousandfold increase. "So that would be like having a baby that's five pounds that will grow up to be four tons," she says. Fortunately for our waistlines, however, our brains consume a good bit of our daily energy budget. And although the octopus is an undeniably quick-witted creature, it isn't bothered with the cognitive demands of wedding plans or strategies to pay for their offsprings' college tuition.

Octopuses' extreme growth rate might also have implications for human biomedical research, Ángel Guerra from Vigo points out. "From a medical point of view, it's very interesting," he says. The hormones and other enzymes that spur on an octopus's growth might some day prove helpful for people in need of growth stimulants.

Within the same species, sizes can be vastly different. A common octopus fella living in the Mediterranean can be ready to find a mate when he is about 6.7 ounces and his body is about 3.7 inches. The gals get a little bigger, reaching about 5.3 inches. Out in the Atlantic, however, the males will bulk up closer to 30 ounces and the females to 44 ounces when they are mature. Researchers presume these differences are largely due to water temperature differences, but they could also have to do with food sources and even pH level.

This quick and efficient growth and short generations make them an appealing animal to try to grow in captivity, like commercial farm-raised fish. "They have these awesome attributes that you would really want in an aquaculture animal," such as their ability to rapidly turn food into muscle mass (which can become food for us), Wood says. But the challenge is their early-life diet. Young octopuses need live food (which can include other young octopuses) to thrive. This makes for labor- and space-intensive work. And live shrimp aren't all that cheap—at least not cheap enough to make the whole process cost-effective. "There is a huge cost in their food when they are small," Wood notes. For some reason, they outgrow this requirement eventually and will start taking to frozen food. "Economically, it's going to be real hard to do," Wood says about commercially rearing them. This is bad enough for the casual octopus consumer, but for research labs and aquariums, it's even more of a bummer, because live octopuses don't ship particularly well, as Wood points out.

Until a booming octopus aquaculture emerges, those interested in studying or eating them (or both) will mostly have to rely on extracting them the old-fashioned way, luring them out of their dens and gardens one at a time.

Secret Gardens

As octopus fishermen will attest, their quarries aren't like schooling cod or even shoaling squid. They're lonesome beasts that live one to a den and have made a niche for themselves by being some of the sneakiest creatures in the seas.

Octopuses choose dens of all types, although their dwelling spaces are not usually much larger than a few times the octopus's own body—cozy undersea studio apartments, if you will. Their hiding is in large part a defensive measure, because they lack the protective hard shell of their ancestors. Most octopuses find shelter in naturally occurring rock outcroppings but can also live in human-made structures, such as old pilings or even shipwrecks. One such wreck is the clipper ship *Warhawk,* which sunk in Discovery Bay near Puget Sound in 1883. It has since become, as biologist Roland Anderson described in the book *Octopus: The Ocean's Intelligent Invertebrate,* "an octopus condominium." He found at least eight giant Pacific octopuses making their home in its ballast pile. Some smaller species have been known to hole up in discarded bottles and cans and even empty clam shells, which they can keep closed from the inside, using their sucker-studded arms. Others octopuses, such as the wunderpus (*Wunderpus photogenicus*) and the mimic octopus (*Thaumoctopus mimicus*), will happily burrow into mud or sand as shelter. Even in small tanks in research labs, octopuses often sequester themselves in whatever enclosure they can find. Scientists and casual aquarists alike have discovered that octopuses are especially fond of flowerpots—they prefer them unglazed and partly overturned (in case you are looking for just the right gift).

When an octopus finds a new den, it might spend some time cleaning it out or doing a little remodeling and redecorating. Despite all of this domestic effort, however, an octopus usually isn't terribly territorial. In fact, it moves more often than the average New Yorker. Once it has thoroughly explored its area for good eats—whether that takes days, weeks, or months—an octopus will relocate to a new neighborhood and again set up shop.

The octopus is not known to undertake the vast migrations of some other sea creatures. Its modes of conveyance aren't especially conducive to long journeys. As a tiny new hatchling, a shallow-water octopus will drift about in the water column for a while, but soon it will settle on the seafloor, where it usually gets around by crawling along the ground. If it needs a quick burst of speed, it can propel itself with a jet of water, which it sucks into its flexible mantle and shoots out through its funnel. This propulsion method is nicely—if simplistically—demonstrated in the classic 1955 octopus horror film *It Came from Beneath the Sea.* In the fictional movie lab, one of the silver-screen scientists blows up a balloon and, without tying the opening shut, lets go of it. The balloon goes flying: "He's jet propelled," the scientist explains of the octopus.

Despite this propulsion option, the octopus is not, on the whole, very speedy. It generally tops out at about twenty-five miles per hour. Even a short sprint at that speed can tucker it out quickly. Hoofing it (or arming it, as it were) along the bottom is even slower going, taking about ten times as long to travel a single foot. But that seems to suit the octopus just fine. It doesn't need too much speed to catch prey. Most prefer to dine on crawling crabs and helpless bivalves anyway.

Master Editors

To live so successfully in so many of the world's diverse ocean climates, octopuses have developed some stunning adaptations, from their three hearts to their blue-hued blood. But some octopuses continue to make further tweaks to their unusual anatomy—some very deep tweaks. We're talking RNA editing, in which the very genetic code is changed to alter the proteins that will be manufactured and how the nerve cells send information.

In our bodies as well as those of octopuses, nerve cells, which control everything from muscle movements to thoughts, depend on electrical signals for communication. To create these signals, they need just the right movement of charged sodium ions and charged potassium ions across the cell membranes. The timing of the sodium and potassium ion movement is critical. But these processes are temperature dependent, which can pose a problem for animals that rely on ambient climes for their own body temperatures. Some octopuses have figured out a way to alter these speeds, allowing them to live in supercold waters. And thanks to these adaptive changes made by polar octopuses, researchers might be a step closer to creating some innovative cures for human genetic diseases.

"It's a cool trick," explains Joshua Rosenthal, of the University of Puerto Rico's Institute for Neurobiology, in his understated surfer manner. "They are using this system to basically rewrite the genetic code of their proteins that are involved in electrical excitability," he says. "All these proteins they are making—tons and tons of slightly different copies through this process of RNA editing.

"It's because they're exotherms," Rosenthal offers as his best guess as to why so many more editing sites have been found in cephalopods and flies than in humans and other mammals, which maintain relatively consistent body temperatures. In cold temperatures, the sodium potassium ion channel communication system that neurons rely on starts to become unbalanced. The potassium ion channels slow down much more than does the sodium side. This could cause some serious problems, but cold-water octopuses seem to be making tweaks to their RNA that speed the potassium channels back up to a more operable pace.

When he was a graduate student, Rosenthal started working with now semifamous Humboldt squid wrangler William Gilly. Gilly was then immersed in the molecular inner workings of the marvelous and obliging giant squid axon. This long nerve fiber allows squid to send rapid signals via sodium-potassium ion exchanges, enabling it to jet rapidly (much faster than an octopus) through the water. We owe much of our understanding of the great complexity of our body's nervous system and electrophysiology to the giant squid axon, which has served as the basis for much of modern neurophysiological research. As a molecular biology graduate student, Rosenthal realized that no one had attempted to actually sequence and clone the squid ion channels that had made the giant axon famous. Back then, in the 1990s, everyone was cloning ion channels from fruit flies and rodents, Rosenthal points out, "so frankly the project didn't turn out to be that interesting." But what did turn out to be interesting was something that at first presented itself as a huge annoyance. No matter how carefully he tried to control his cloning process, he could never quite get the genetic sequences of his clones to match up entirely. "There were always a couple of points that were variable and either had an A or G." So, naturally, that suggested there was a problem in the sequencing. "Then I go back and clone another. I clone *another* one, and I'd always see that variation at this certain place."

That predictable variance turned out to be a site of RNA editing inside the cells. "But, you know, it takes a while to figure this stuff out," he says humbly. It was only during his postdoctoral research that he started looking into octopuses and took up the more interesting investigation of *which* sites were being edited and what they were doing to the channel's function. This RNA editing had been described in humans and other mammals, such as rats, but Rosenthal stumbled across it "happening in orders of magnitude more frequently in octopus than any other organism." After searching tens of millions of sites in the human code, for example, scientists have found only a few dozen editing sites. In octopuses and squid, Rosenthal and his graduate student Sandra Garrett have found more than a hundred RNA editing sites in a little more than ten messenger RNAs. "I tell you, these things are machines for editing," he says of the octopuses. "They are unlike any other organism."

So now the question is whether the enzymes that the octopuses are using to do all of this editing have evolved specially in these animals. He points out that because we have found only a few dozen editing sites in humans and there are hundreds or perhaps thousands in octopuses and other cephalopods, the enzymes that are in charge of the editing will be interesting to study as examples of divergent evolution. "These enzymes are capable of editing at very high levels through structural modifications that have evolved over millions of years," Rosenthal explains. To do this, cephalopods have been adding molecular domains that other lower mollusks don't seem to have. But little research has gone into figuring out which other animals—particularly cold-blooded ones—might be taking advantage of this tool. So whether reptiles and amphibians are doing it at a similarly high level is currently anybody's guess, he says.

Rosenthal would also like to figure out whether these changes can be made by individual octopuses as the temperature changes or whether they are more long-term adaptations of a whole species. If it's the latter, it might be thousands or millions of years in the making. But there's a chance it's the former strategy. That would mean that an octopus could switch by the season—or maybe even by the hour. There will have to be new experiments to find out, because the octopuses that they used in their earlier study came from pretty stable temperatures, such as those in the Antarctic.

But simple observations from the field suggest some octopuses are at least doing *something* to adjust to large temperature swings. The Mexican pygmy octopus (*Octopus digueti*) are little guys that Rosenthal describes as about the size of a gum ball. They often take up residence in whelk shells in the estuaries of Baja California. The water there can be up around 90 degrees Fahrenheit during the summer and then drop down to the forties at night in the winter. A single octopus would likely live through the better part of the temperature swings, which suggests they might be using RNA editing to adjust to such vastly different ambient temps.

Temperature is just one variable an octopus confronts. These animals have been able to flourish in many extreme environments and adapt to very different living conditions using very similar genes. "They might be regulating their editing for lots of different reasons—food, light levels for vision," or others, Rosenthal says. Antarctic octopuses confront extreme darkness, especially during the sunless season, so transient RNA editing might be boosting vision or other sensory abilities. Life experience, such as the amount of stimulation and the opportunities for learning and memory creation, might also influence RNA editing, Rosenthal hazards. Octopuses and other cephalopods "have really taken advantage of this process to help create molecular complexity."

What could this sort of strange action be doing in other animals? The more animal genomes we decode, the more we realize how much we all have in common, which begs the question: Why *are* we so different from the earthworm? RNA editing is one of the ways to manipulate the given set of genes. In this way, it can help create nuance in the genetic landscape to allow different genes to be expressed rather than having to depend on longer periods of evolution to establish whole new genes.

As cool as all of this is to find in cephalopods, the eventual goal is to learn how the editing process works—and to take control of it. "There is a lot more to be learned from looking at the editing enzymes," Rosenthal says. He is interested in "using these ideas to see if we can develop a system where we can guide the editing" to change, for example an A to a G in an organism's RNA code. That could have huge implications, he says, if they can translate it into humans. "I think that will give us a very good idea of how to manipulate human enzymes down the line."

Some diseases, such as some types of cystic fibrosis, are caused by a single genetic point mutation. Because these conditions are triggered by an incorrect sequence in the genetic code, they have long been the target of gene therapy. For this remedy, a healthy genetic sequence is fed into an inactivated virus that then can infiltrate a patient's cells, providing a correct copy of the gene. But this is simply adding another copy.

"If you could *change* that A back to G in the RNA, you could cure the disease," Rosenthal says. Currently he and his team are trying to learn from the octopus's RNA-editing trick how to manipulate the editing enzymes. "If you can make this happen, it will give you a lot more power than traditional gene therapy," he says. One of the researchers in his lab, Maria Fernanda Montiel, is working on finding a way to redirect RNA editing to locations that we would want to alter to cure human diseases.

Rosenthal reassures me that instead of injecting octopus enzymes, they would start with the human version and work backward "to manipulate it to behave like the octopus one does in a certain context." This would make it more readily accepted by the human body as well as by the Food and Drug Administration. "The enzyme that does the editing is structurally quite similar" to ours; the octopus has just tweaked theirs a bit, he says. "So figuring out that tuning, how their activity is slightly changed, is important."

RNA offers the other advantage of being accessible. Unlike DNA, which is hidden away in the cell nucleus and guarded, like a family secret, RNA is floating around in the cells' cytoplasm. The only trick with RNA is that it is continuously refreshing, so edits to its code need to be made over and over again.

"I think the potential for this is tremendous, and really what we've learned from all these other organisms is incredibly useful," Rosenthal says.

Indeed, even as the octopus might seem so utterly strange to us, with its many flexible limbs, three hearts, blue blood, and amazing molecular editing, it might just provide us with lessons for our own health as well as insights into the evolution of complex organisms the world over. We just have to be open-minded enough to know where to look. And with the octopus, that can be challenging, because they are the ultimate masters of disguise.

CHAPTER FOUR

Skin Tricks

An octopus might look as if it would be a bit slimy to the touch, but its slick skin is one of its most amazing assets, allowing it to blend seamlessly into the background in color, brightness, and even texture and movement.

Plenty of other ocean animals have countershading. They appear light from below to blend into the lighter surface if you're looking up and dark from above to disappear into the darker waters below—witness the shark's white belly and darker dorsal area.

But the octopus has gone a gazillion steps beyond this static tactic. Hundreds of millions of years in the making, its color-changing capabilities had no analog in the human world until the advent of colorized movies and television just last century. And even those are but a superficial flourish compared to the amazing mutability of octopuses, cuttlefish, and squid. Teams of researchers and millions of dollars have not yet been able to fully understand or even begin to replicate it.

Some species of lizards can of course change the color of their skin. But unlike the four-legged reptiles, which shift shades in a matter of seconds, octopuses can complete a full transformation faster than a magician's assistant can change the color of her dress—and without any smoke, mirrors, or curtains of ink. A full-body change can happen in about three tenths of a second.

And that's just the beginning for the octopus. It makes the chameleon's pallor-changing abilities seem no more than a paltry parlor trick—adding reflectivity, texture, and even light to its pellicle palate.

This charade has been borrowed to outfit some sneaky comic book characters. Will Eisner's mid-twentieth-century villain the Octopus—

who is enemy to the Spirit—is a renowned master of disguise. He blends into his background so well that often only his white gloves are visible. And in the flashier 2008 film version, with Samuel L. Jackson taking the role, the Octopus appears in different costumes (from Nazi to samurai) to match his various exotic settings.

These big shifts are one reason it has been so difficult to identify and describe different octopus species. Trying to identify a bird out in the field can be a challenge. But at least there are hard-and-fast details to check against a guidebook—a spot of white on the wing, a curve of a beak, a notch in the tail. With an octopus, you're dealing with changeable color, pattern, *and* shape. Octopus skin is so variable that it can make a wild octopus hard to identify not just as a particular species, but even as an octopus—and not a rock, coral, plant, or even a fish.

These mind-blowing abilities have caught the attention not just of biologists, but also of researchers as far afield as mathematics and nanoengineering. Richard Baraniuk, a signal processing professor at Rice University, had never cultivated much of an interest in cephalopods. "I had no idea how cool they were," he tells me in his animated voice one afternoon when I join him for coffee at a café on campus in Houston.

Baraniuk is working with a team of other scientists at Rice, in collaboration with folks at Woods Hole and other institutions, to better understand cephalopod skin so that we humble humans might someday be able to mimic *its* amazing mimicking capabilities.

Without the benefit of LCD (liquid crystal display) or other technology, Mother Nature has wired the octopus with a high-definition display of millions of chromatophores, tiny color-filled sacs in its skin. Nearby muscles expand or contract the chromatophores based on signals that, not unlike LCDs, tell them to activate certain hues. The sum effect of these color-containing packets determines the octopus's overall shade and patterning. Researchers have even found that among the octopus's turning tricks, they can reflect polarized light—a type of wave oscillation we can detect only with special glasses but that many ocean-dwelling organisms can see just fine.

Octopus exhibiting rough skin texture via papillae.
(Denise Whatley/TONMO)

In addition to chromatophores, octopus skin has iridophores, which are shimmery reflective cells that can give off a variety of hues. It also contains leucophores, which are static white elements that help enhance the rest of the camouflage display. And below all of this, it boasts a fine-grain network of muscles that can relax or flex to create bumps, or papillae, to match its three-dimensional environment. Not too bad for a mere mollusk.

All of this flair is not just for showing off. So what are they using it for? Decades ago, some researchers thought cephalopods might be using it as a sort of visual language of the skin, says Jennifer Mather, a biologist at the University of Lethbridge, in Canada, who has studied cephalopods for more than four decades. But now we know better.

The octopus is not particularly social, so it isn't likely employing its skin to send secret messages to a cephalopodian confidant. Even squid, which do travel in shoals, don't seem to have any serious skin-based conversations. Squid, cuttlefish, and even some octopuses have skin colors or patterns specific to courtship and fighting; but considering their full range of color-changing skills, these are pretty simple—a flash of deep red or white, for example.

"Clearly they *have* the ability to generate more complex communications with skin displays," Mather says of octopuses, squid, and cuttlefish. "I think they haven't done it yet because they don't have anything complex or interesting to say." Despite the large groups of squid that often swim together, she notes, they don't seem to be interacting in nuanced ways like a pack of wolves or a herd of elephants. "They're just hanging around together." And sometimes they're even cannibalistic, so a witty repartee or squid soliloquy shown on the skin isn't likely to serve them so well evolutionarily.

Despite all of the octopus's flashy flesh, however, as far as we know, it is color-blind. And we don't know if it is even *aware* of what its skin is doing at any given moment. So how has it evolved such sophisticated color-matching techniques?

Other animals' vision has probably been the driving force behind the octopus's camouflage. By observing what it does with its skin, we are learning that many of its tricks are part of an elaborate strategy to avoid becoming lunch (with the side benefit of helping it catch its own).

In the animal world, eyes equal animal—which often equals food. Some octopuses can produce temporary ocelli, which are fake eye spots like the ones some species of butterflies have on their wings or some tropical fish have on their tails. These two big spots appear on its body, perhaps to confuse predators into lunging at its tail end, giving the octopus a better chance to jet away from the miscalculated attack.

Another lesson learned over the eons is that big animals tend to have big eyes. And although an octopus might not *know* that, it takes advantage of this natural trend to psych out predators. Some octopuses put up a good bluff by forming big dark circles around their real eyes to suggest that they're much larger than they really are.

An octopus can also try to camouflage its own eyes, a tactic that ap-

A common octopus hides in a hole while displaying a large dark patch around its eye.
(Denise Whatley/TONMO)

parently works well, because a large percentage of nerves that control body color are dedicated to the skin around the eye area. Eye camouflage alone can call on 5 million chromatophores, points out Roger Hanlon, of Woods Hole. The octopus can also produce dark horizontal bands—called eye bars—alongside its eyes to obscure them in a larger, noneyelike shape. Or it can try to disguise its eyes by sprouting little temporary horns via the muscle-controlled skin above the eyes. This little disguise along with the ability to turn a deep fiery red once earned the octopus the nickname "devilfish."

The octopus also calls on certain color-changing tricks to be a better predator. Like many hunters, the octopus relies in part on motion to spot a potential meal. If a skittering crab starts to get suspicious that it is being tracked, it might freeze to avoid detection. But some researchers speculate that crafty octopus species will hover near a crab and create a "passing cloud" display. In this skin show, a dark spot forms in the middle of a lighter patch, and the little composition moves rapidly across the octopus's body. This effort might startle or confuse the prey into moving so the octopus can spot it and snatch it up without its having to move and risk losing track of the target. Diabolical.

Science has shown that some of the octopus's color changing can indeed be more direct communication—but those signals are often aimed toward other species. One infamous example is the blue-ringed octopus genus *Hapalochlaena*. The blue rings all over its body flash in warning when it feels threatened (which is considerate, given its deadly, poisonous bite). Even the two-spot octopus (*Octopus bimaculatus*) will flash a blue ring when it's agitated, as one at the Woods Hole lab did toward me when I crept up to its tank to take a picture.

A glowing Technicolor (and techni-texture) dreamcoat or David Bowie–esque eye ornamentation, however, is but a facile strategy in some octopuses' bag of tricks. Resembling a rock is all well and good until you need to make a getaway. But how do you do that under the gaze of a hungry shark without becoming a *dead* giveaway?

Easy: Look like something other than a soft, tasty octopus.

Many fish are thought to hunt using a visual search image—a mental picture of what their target looks like. So if all of a sudden the easily identifiable orangeish octopus has turned into a white blob, it can throw off the assailant in time for the octopus to get away and camouflage against the seafloor.

In perhaps the most famous octopus video clip, used in ocean explorer David Gallo's 2007 TED (Technology, Entertainment, Design) talk, a common octopus near Grand Cayman is excellently camouflaged on a clump of algae. Startled by the underwater human diver, it suddenly flashes white, then shoots out a cloud of ink as it jets away. Still pursued by the videographer, it lands on the seafloor and spreads its web wide, trying to make itself look as big as possible, with big dark circles around its eyes to make them look much larger. The short clip, filmed by Hanlon, elicited gasps of astonishment from the TED audience when it was played backward in slow motion so that they could see that the quick-change camouflaging was no computer-generated trick.

But Hanlon himself wasn't as shocked as the uninitiated viewer. "They can camouflage anywhere they go—and they can do that in a fraction of a second," he says, when I meet him in Woods Hole on a chilly clear November day. Hanlon's office is packed with books from his decades of study. Almost before I can sit down he launches into a lecture. Hanlon's occasional impatience is perhaps well earned. He has spent countless hours underwater in oceans and seas all over the world watching and often filming octopuses going about their quietly astounding lives. He has hundreds of these hours on video and some forty thousand still images captured from all over the globe.

Sometimes he simply tries to observe the octopuses while he remains unobserved by them. Other times, he gets involved to study their reactions. For one paper, he was studying defensive coloring, so, he recounts, "I became the predator, and I 'attacked' them with the camera." With repeated attacks, he was able to capture—and then quantify—the animals' responses. "When the camouflage fails," he says, "they go to the secondary defenses."

A secondary defense mechanism—after an octopus realizes it has been seen—can include shooting out ink or spreading out to look as large as possible, as the one in Hanlon's famous video did. The goal, he says, is "to make the predator just hesitate for a moment longer," affording the octopus more time to make an escape.

In one video from Hanlon's lab, an octopus can be seen traversing a seafloor convincingly disguised as a plant-covered rock. "It's taking advantage of the 3-D structures," Hanlon explains, pointing to the disguised octopus. As the octopus moves slowly across the floor, it keeps its form perfectly, raising some of its arms—now bumpy and greenish—up in the air to mimic plant life. As a result it needs to "do this rather athletic two- or four-armed walking backward, holding everything in place," Hanlon notes. The octopus looks a little like a flamenco dancer holding its arms up with a flourish. "People call this the flamboyant pattern," he says. With its skin papillae raised, and in its shades of brown and dusty green, I never would have picked it out for an octopus, flamboyant or not.

Not only is the octopus well disguised, "he knows where he is going," Hanlon notes. "He is heading for that rock, right there," he says, pointing to a rock in the distance. So the octopus moseys along quietly and pauses at one rock, for which it's a pretty close match. "He's got pretty good camo anyway, but he didn't like that," Hanlon says. So instead, the octopus takes off for a more distant destination: "The little bugger goes out here," Hanlon says, pointing to a rock farther away, near the murky horizon. "So he is deciding where he is going and how he is going to get there."

Hanlon isn't interested in characterizing this impressive shape-shifting display just on superficial appearances. He also wants to examine how fast a camouflaging octopus moves. The ocean is a dynamic place, where things are never quite still. Some octopuses have really mastered this aspect and can let their arms or bodies wave in the current just like a sprig of seaweed. Hanlon shows me another video, which was filmed near Vigo, Spain. The area might be ideal for an octopus, but for diving, he says, it's "deep, dark, and dangerous." (He's cagey about exact locations, almost like the Spanish fishermen. They're "one of my best kept secrets—my dive spots," he says.) The shot opens on an animal that is hunkered down on the seafloor in full camouflage mode. "You just can't believe it's anything other than a rock," Hanlon says. Then it starts to move. "It does this moving rock trick, where it keeps this posture and is moving across the way here. So here it comes, perambulating along, and it's heading toward this kelp, and—blink!—now, not only did it change its pattern, it's changed the color, and it even changed the papillae when it got here." In less than a flash, the octopus has gone from looking like a rock to being camouflaged against the kelp, completely disappearing into it.

Looking closely at a video of a camouflaging octopus, Hanlon points out the well-hidden octopus and then asks a rather surprising question: "Does that animal look *exactly* like any other thing close around it?" No, in fact, it didn't look precisely like anything behind it.

"That is the secret—this is the magical thing," he says. "Everyone has this ridiculous misconception" that the octopus creates an exact match. There is almost no way the octopus could look *exactly* like its environment, he says. On a given coral reef, there might be hundreds of different species of corals, in addition to sponges and other species. "Can you look like five hundred to six hundred things? Ridiculous, right?"

Hanlon suggests that octopuses, as well as other animals, have a limited repertoire of different patterns from which they can draw. By studying camouflage patterns across the animal kingdom, he and his colleagues have established just a few types: uniform (a consistent coloration), mottle (multicolored), and disruptive (bold markings that hide the animal's own outline). The researchers have developed digital imaging programs to transform the pattern into plots contrasting light and dark spots. That provides "a somewhat quantitative measure of what the pattern really is," he explains. Octopuses (of course!) can do each of these. They, along with their cephalopod cousins, might be the only ones that can engage in all of the tactics.

So by Hanlon's reasoning, an octopus will often look *like* something next to it. But to reach that final appearance, it will draw on its familiar patterns and the cues it gets from its environment to blend in just well enough to slip by the predator's perception—or at least to keep from being recognized as food.

More recent findings have shown that octopuses are, in fact, picking and choosing elements of their immediate environment to match rather than to blend in with the whole area. In a 2012 study, a team of researchers at Ben-Gurion University in Israel used a software algorithm to analyze colors, brightness, and patterns of common and day octopuses (*Octopus cyanea*) camouflaging in their natural habitats. They found that instead of resembling an average of the whole visual field, the octopuses actually more closely matched select landmarks, such as a piece of coral or a clump of algae.

This tactic, the researchers posited, helped to address one of the big problems in camouflaging, what they called the point-of-view predicament. For example, if you put all of your effort into making yourself invisible against the sandy bottom, but only to a shark swimming far above, you will likely be spotted quickly by a hungry fish passing by at your level. But by essentially becoming a new particular feature in that landscape—a rock, piece of coral, or a plant—an octopus can fool predators and prey coming from almost any direction.

Octopuses, of course, are not the only ones to have developed clever disguises. Several organisms have evolved to resemble less palatable species—take the famous viceroy butterfly, which looks similar to the more toxic-tasting monarch, or the South American catfish species that finds safety in numbers by evolving to look like local spiny cousins. But some octopus species not only adopt the coloration of other less appetizing animals, they can also contort their flexible bodies to look—and move—like them.

The mimic octopus (*Thaumoctopus mimicus*), which lives in the shallow sea bottoms of Southeast Asia, does not seem a great candidate for masterful disguise. It has long, tendril-like arms, and its skin has a bold stripe or spotted pattern, making it stand out against the drab sand below. But its skills are in its moves. These mimic octopuses, described formally in a paper by Mark Norman and Eric Hochberg in 1998, seemingly impersonate a rather stunning variety of other less lunch-worthy animals. They have been spotted flattening out and swimming along the bottom like a flounder or extending two opposite arms while hiding the other six and most of their body in the sand to appear like a banded sea snake. Other muses for this octopus appear to include the poisonous lion snake, stinging sea anemone, and just plain unpalatable jellyfish. And although the knowing human viewer might be able to pick out the octopus from the actual lion snake, to a predator, the former might at least look risky enough to pass up.

Watching these transformations on video is even more uncanny than watching a trained sea lion give a high-five or a dog fetch a specific toy by name. These tricks of the octopus kingdom are somehow established on the go, whether through evolution or observation—not in captivity.

But we are not the intended audience for these charades. In 2011, Hanlon and his colleagues published a paper describing a new visualizing technology called hyperspectral imaging, which can show us what a camouflaged octopus looks like in the eyes of potential fish predators. Hanlon calls their new tool "the world's fanciest camera." It doesn't just record the red-green-blue colors that we humans are best tuned to perceive, it also captures a much wider range of wavelengths in each pixel of the digital image. With this wealth of data, researchers can then filter each image to what we know about the visual range of the predator "to see what the predator sees—and not what the goof human sees," he says. The findings reinforce for Hanlon the idea that "we've got to see what the world is seeing out there and pay attention to the visual systems." We fancy ourselves pretty good at visual perception, "but that's just human ignorance," he says. For example, "we can't see polarized light; we can't see anything under dark conditions; we don't see anything in UV. And there are animals that have all those capabilities"—and more. So as impressive as octopus disguises might look to us, they are actually "tuned to all those predators," Hanlon says. And if we can visualize what the predators see, we will be able to much better understand the true magnificence of octopus subterfuge.

Illuminating Disguise

Color and texture—and even mimicry—might be some of the more obvious disguises in the octopus's closet. But it also has mastered some far subtler strategies to truly vanish before your eyes. Imagine you wanted to disappear against your floor (perhaps you've just done something terribly embarrassing—or a giant barracuda is after you). The floor's appearance is made up of much more than just segments from the color wheel or textures from a touch-and-feel kid's book. A well-waxed wood floor, for example, has a certain luminosity to it. Even the most high-quality photo of the floor would have a hard time vanishing into the real thing in natural light. So to really blend into a background, you need to match its brightness in addition to its colors and texture.

And the cool thing, says Richard Baraniuk, is that octopuses as well as squid and cuttlefish manage to do this without actually producing light. So, as he explains, it's like the difference between an e-ink Kindle and an iPad. The Kindle passively relies on the ambient light to make its display readable, much like a real piece of printed newspaper or a page of a paper book, whereas the iPad must actively produce light in order to be read, like a computer or a television display. Octopuses, squid, and cuttlefish are rather more like Kindles in that the animals are using ambient light for their display, Baraniuk notes. This is the strategy, of course, that most other animals—from cheetahs to katydids—use. Actual light creation is relatively rare in the natural world, being limited to animals that produce a true light-generating glow, called bioluminescence. The octopus's ability is impressive in that it is using just passive light to create an active, constantly changing display.

To truly go unseen, octopuses cannot simply match a pattern or color in their environment. As we sit in the warm, sunny Texas afternoon, Richard Baraniuk explains this phenomenon another way. He suggests a disappearing trick worthy of a magician's prime-time television finale: making a semi truck vanish—from front grille to trucker girl mud flaps. You could paint it expertly with colors to match the sky and the grass and the clouds. Heck, let's say you could even invent a smart paint that moved so that the scene changed according to the viewer's perspective—and to pull from the octopus's bag of tricks, let a few painted passing clouds float by along its surface just to dispel any doubt. But there would still be a huge flaw in that design, despite all of the engineering effort. We would be able to spot right away where that truck started and stopped, Baraniuk says. How?

The truck is parked in the bright Texas sun, so to disappear in that environment, it, too, must somehow be really bright. Otherwise, "just the fact that it's a big three-dimensional object means it's blocking out a lot of light," Baraniuk explains. And that would ruin the illusion.

Suddenly the problem is not as simple as it sounds. But the octopus has it under control. Nature has outfitted this animal with iridophores, reflective cells that are distributed throughout the skin. These reflective patches help immensely in allowing the octopus to disappear without having to develop an LCD-type glow, and thus be more like the e-ink Kindle than the iPad.

This light use can be breathtakingly subtle. Lydia Mäthger, working in Hanlon's lab at Woods Hole, found that at night, cephalopods can even camouflage themselves under light levels similar to that of starlight (0.0003 lux; daylight, for comparison, is closer to a range of 10,000 to more than 100,000 lux).

Mäthger, whose shared office overlooks a boat-filled inlet on Cape Cod, is a slim, energetic German woman who is keen to explain these nuanced structures that so fascinate her. Iridophores are especially interesting because they are entirely different from chromatophores, she notes. While chromatophores are tiny, they are still much larger than iridophores, which are about one hundred microns (each micron being one millionth of a meter). So, we're talking structures ten thousand times smaller than a meter, smaller even than the width of a human hair.

Unlike chromatophores, iridophores don't have one single color. But that doesn't mean they only appear as a piercing clear or a slicing silver. These cells create color through a process known as "structural interference," which, as Mäthger explains, is "essentially like a soap bubble." A bubble will appear shiny with color reflected on its surface because its film is just the right thickness to reflect certain wavelengths of light, thus its structure interferes with a particular wavelength. So a thin plate will reflect a different wavelength than a thick one. "They basically tune into the wavelength that they reflect" by giving the iridophore plates more or less thickness, she explains. (All of this is thanks to protein changes, but we don't have to get into that.) In cephalopods the iridophores can actively change their reflectance from reflecting red to yellow and green.

But let's not stop there! Iridophores also reflect polarized light. In fact, Mäthger notes, "iridophores are just some of the best polarizing structures out there." The iridophores reflect polarized light at the same angle as it was received, she and Hanlon have found in their research. Cephalopods might be using this ability for communication, because the polarized light reflected by iridophores is stronger than the polarized light conditions of the environment around them. This idea is still part speculation, but if these animals do use these specialized abilities for a communication channel, they might, for example, use a flash of ultrapolarized light as a warning sign to other species that can also see that type of light.

A third player in this dancing color drama is the leucophore, which provides a "beautiful white base layer," Mäthger says. These protein-based structures set the backdrop for the more dramatic colors and reflectors. As far as we know, leucophores are passive and don't change or shift in the way that expanding and contracting chromatophores or thinning and thickening iridophores do, Mäthger notes.

Some unusual cephalopods' skin has also been shown to contain photophores, structures that actually emit light via a chemical reaction. True bioluminescence is known in only a handful of organisms, including fireflies and some species of jellyfish and plankton. Squid are more commonly known to flaunt this ability than octopus species. But one species of deep-sea octopus, *Stauroteuthis syrtensis,* is able to emit light from its suckers. According to researchers who handled it, these suckers had lost their suctioning ability and become full-time, light-producing photophores. A report about it published in *The Biological Bulletin* describes the luminous suckers that "either glowed dimly and continuously or flashed on and off more brightly." These flares in the dark might be used to scare off predators and to lure in prey.

Mäthger takes me to the lab where they keep their cephalopods. They currently have only one octopus in residence. It's a two-spot octopus, which is suctioned magnificently into a corner of the small tank, providing us a view of its amazing collection of white suckers from one side and its elegant brown body from another. As I approach the tank it narrows its eyes to slits and starts flashing a slender blue ring right above the funnel at me, to show it's annoyed—and that it doesn't know me, Mäthger says.

Even Mäthger, who has spent years looking at these flashy creatures, is still awed by their full display. "That's mind-boggling to me how that works—that's something that we're still trying to figure out."

Woods Hole octopus flashing its blue ring at me.
(Katherine Harmon Courage)

By studying these impressive light-bending abilities, researchers hope to learn more about how these animals are able to pull off such amazing vanishing acts.

Controlling Colors

The seemingly instantaneous change of an octopus's camouflage can appear miraculously fast. Especially considering that each little chromatophore is being controlled by muscles that must push or pull in a perfect choreography to get the correct overall display.

But the muscles are the easy part, Richard Baraniuk says. They're part of a nice macrophysical system that you can get an easy conceptual grasp of. Behind those muscles, though, a web of more mysterious nervous system controls are pulling the strings, so to speak.

These signals come from neurotransmitter compounds. A little spurt of the neurotransmitter serotonin (a version of the one that flows through our nervous system) causes the muscles around a chromatophore to relax, allowing the pigment area to expand. And then when the muscle contracts again, the color spot shrinks away.

That is likely just the beginning of how the octopus's complex camouflage is being controlled by these rapid, fluid signals. By learning more about how neurotransmitters control behaviors in the octopus, we might also discover quite a bit about how they're operating elsewhere in the animal kingdom—including in us.

"There are so many different types of chemical neurotransmitters, we don't even know how many—and we don't know what they do," says Naomi Halas, a professor of electrical and computer engineering, biomedical engineering, chemistry, physics, and astronomy at Rice University, where she is also the director of the Laboratory for Nanophotonics (and of not slacking off). Chemical signals are difficult to decipher in part because there seem to be so many of them, and because we have only just begun to appreciate their role in the nervous system—both in animals' and in our own. In fact, these signals are going to be to the next several decades what the electric nerve pulses—the charged signals that help to cue our muscles and thoughts—were to the past several, she says. "This is going to be, in the next fifty years, a huge thing," Halas says. Research has shown, for example, that these neurotransmitters, such as serotonin, not only affect daily mood and sleep, but abnormal levels of them seem also to be involved in diseases, such as Parkinson's.

So taking a purely electric view of things is missing a big part of the neurological picture—both for octopuses and for humans. "I don't understand how people can develop sophisticated neural engineering models and completely neglect the chemistry—and expect them to be in any way comprehensive," Halas says. She hopes that lessons from the octopus will help us "think again about neurotransmitters." And from studying these chemicals in octopuses, "maybe we can learn more about neurotransmitters in human beings," she says.

Researchers are now taking some of the octopus neurotransmitters for a test drive to see what they can do. Roger Hanlon recently found that he could stimulate chromatophores using acetylcholine, a neurotransmitter that activates muscles. Control of iridophores has been even more difficult to pin down. At Hanlon's lab, Mäthger has been able to show that iridophores are also being controlled by the nervous system. The curious thing is that she and her colleagues haven't yet found any nerves that seem to be controlling iridophores directly. They have found one that's coming in nearby to make a synaptic connection. And they can show that the neurotransmitter acetylcholine is being fired in the area. But then they lose track of the signal. The control of the iridophore remains mysterious.

Despite everything we are learning about the muscle work and pumping of neurotransmitters, scientists are still not sure how energy intensive camouflaging is. Some octopuses, including ones Hanlon has found, seem to change their appearance almost excessively. He has documented one octopus on a Pacific coral reef changing colors and body patterns 177 times in one hour. That seems like it could get rather exhausting, moving not only the millions of chromatophore muscles but also those under the skin to change texture. But in the dangerous ocean, perhaps it is best to err on the side of safety and stay well hidden. Continuing research into how the octopus is ultimately manipulating its chromatophores and iridophores will help biologists understand how they are able to complete their transformations so frequently, so quickly, and so stunningly.

"Eyes" in Their Skin

All of the quick camouflaging begs the question of how an apparently color-blind octopus so accurately changes to resemble aspects of its background. Some researchers propose that the octopus might not just be using its eyes, but that it might also be able to "see" with its skin.

Hanlon and his colleagues have discovered that cephalopod skin contains opsins, which are light-sensitive proteins found in the eyes of many animals, including cephalopods and us. In the eyes, these molecules detect photons of light coming in and produce a corresponding signal to send to the brain. There are several different kinds in our eyes alone, some of which are involved in "vision" as we know it and others that help to regulate pupil size or circadian cycles. But scientists are still studying how they work in humans, so deciphering their function outside the eye—and in invertebrates—is a whole other biochemical can of worms.

"We were very excited to find opsins expressed in the skin," Lydia Mäthger says. "The obvious question that we asked immediately was: 'Can that help camouflage?'"

As her colleague Hanlon explains, opsins in the skin might sound exotic, but there is perhaps little difference between them and other sensors that we have in our own skin, say for temperature. Regardless of what information the sensor is gathering, whether it's temperature or light, it is likely turning that input into a nerve signal. "And that neural signal is going to send the information right up my arm and into the old melon," Hanlon says, pointing to his head. In our case, "then the brain is going to send that information back out and tell the body what to do."

That scenario is the typical solution across the more advanced members of the animal kingdom. But octopuses, as we know by now, are not exactly typical. "We think there is a good possibility that there is another option here," Hanlon says. "And that option is that it senses the information out there in the skin and does *not* send it back to the brain, but rather uses it locally." A 2011 study showed that we also have at least some photoreceptors in our skin that sense light. Called rhodopsin, these proteins are also found in our eyes. They can detect UV light, which allows the skin to start getting ready to repair cellular damage wrought by the sun's rays.

If the octopus uses its local skin-based receptors for light sensing, it would still have some sort of neural-sensory pathway, but this would mostly be for the one small area of the skin. "So now you take the brain out of the picture," Hanlon says. And that, he notes, "gives you redundancy and takes a little burden off the brain." All good things in the catch-as-catch-can ocean.

But investigating this ability much further has been a challenge. One way to tell if octopuses are picking up on light and color cues with their skin would be to expose them to a visual cue where their eyes can't see it. But with those big eyes and that slippery, slinking body, "that's been very, very difficult," Mäthger says. "How would you stop an animal from seeing it?" Blindfold. Of course!

But wait—how do you blindfold an octopus? Exactly.

"Yeah, try that," Mäthger says with a laugh. They have "eight arms to rip anything off that you put on them." She and her colleagues tried all sorts of strategies, including Lycra fabric and even goggles. But once the cephalopod knows that you're up to something, it will remove the offending object and make its displeasure known—often by inking on its experimenter.

So to take a more roundabout (and less ink-soaked) route, some researchers have been looking at what the eyes and the skin opsins are each best at detecting. This approach should help illuminate how octopuses and other cephalopods might be using each to coordinate camouflage.

Octopus eyes—as we shall see later—are amazing. They have focusing lenses just like our own, and in some respects are even better than ours (spoiler alert: no blind spot!). The eye is good at providing super-high-resolution detail about the octopus's surroundings. But skin opsins might be feeding the octopus a different type of information about the visual environment. The opsins have no focusing lenses, no complex rod-and-cone makeup. They're basically just receptors. So Mäthger and others are doing research to figure out just what the octopuses could sense through their skin opsins. "We don't have any answers yet," she says.

But they're still also hard at work trying to solve the color question. The big mystery so far is why scientists have found only one type of opsin in the skin. As Mäthger explains, with just one sort of opsin, there is no way that the octopus could discern one color from another "because you need *two* types of photoreceptors to discriminate different colors. Based on one, you can't do it," she says. Once an additional color receptor is present, a larger range of colors becomes visible. (Our own vision is based on just three receptors: red, green, and blue.)

There *might*, however, be a way the octopuses could be doing it with just one type of opsin (although researchers are looking intently for other opsins). The octopus could use its color-changing elements, chromatophores or iridophores, to work as color filters over the opsins. Some birds employ this strategy in their eyes by secreting oil droplets that have a photosensitive pigment, which helps to filter color to give their own color vision a boost. "So the same thing could be happening with the cephalopods," Mäthger says.

Hanlon adds a bit more hope for this hypothesis in noting that the opsins are "not just spread willy-nilly all over the place." Instead they seem to be found in concentrations near yellow-pigmented chromatophores. He, coyly, won't divulge any more detail but notes that it is "a wonderful research lead."

The possibility remains that the opsins aren't contributing to color camouflage at all. A more secure speculation, Hanlon notes, is that they are there simply to detect general brightness, which could be used to cue the luminosity of a disguise. Hanlon and his group at Woods Hole are diving into more opsin research. "We're just beginning to seriously look at how it works."

For his part, Richard Baraniuk is taking a systems engineer's approach to studying how opsins might be involved in the camouflaging process. He calls these sensors a "distributed, lens-less camera." "Their body is a camera. Isn't that cool?" Yeah, or profoundly humbling.

Figuring out the density of these opsins can provide an insight into how high the animal's "resolution" of perception is. That, in turn, can help shed light, so to speak, on how much information the animal is getting—and possibly how it's using it for things like blending in, Baraniuk explains.

But color and light, of course, aren't the only tricks an octopus deploys when it goes into full camouflage mode. Texture presents an additional wrinkle in the visual perception question. By way of example, Baraniuk gestures to the green lawn next to us. If you were to put an octopus down on the ground and it camouflaged itself (although if it were plopped down on a grassy quad in Houston, it would probably be too pissed off to do anything more than turn an angry beet red or blanch into a bright white), it wouldn't just turn green or pick up on the right light levels. It would also activate its subcutaneous muscles, prickling up its skin so as to match or at least more closely resemble the texture of the grass.

While octopuses—and cuttlefish and squid—might be able to pick up light and even color cues from opsins in their skin, these simple sensors are likely not up to the task of picking out three-dimensional texture. For that, they're probably using their eyes, Baraniuk concedes. They might even be calling on a library of different textures in their catalogue, akin to Hanlon's proposal for four hue-based camouflage options. So using this array of textures, they might be able to quickly pick the one best suited to the background on which they find themselves—smooth sand or ripply coral or fuzzy seaweed.

All of these gaps in our knowledge present a delightfully complex project for those, like Baraniuk and Hanlon, who are now looking for new ways to understand how the octopus assimilates all of the detailed information it absorbs from its environment in order to make such convincing camouflage.

Processing Power

There is an awful lot that octopuses need to "know" to blend into their environment. The process of detecting and mimicking color is, of course, quite amazing. And adding light to the equation, Baraniuk notes, "adds an extra degree of complexity." To vanish against coral, rock, or sand, the animal must ascertain how the light is filtering down through the water from the sun—or moon or stars—and bouncing off the surfaces around it. That means a heckuva lot more info than just picking shades of gray—or green.

The researchers have figured out how to mimic this process, somewhat, on a computer. And they are working on algorithms to find the most efficient way to sense light and control output at the same time. This should give them insight into how octopuses might be doing it, in addition to helping us humans accomplish it with our own technologies. After all, with hundreds of millions of years and almost as many generations, Nature has had the chance to get pretty close to optimality.

Baraniuk spent much of his time as a graduate student thinking about something far less luminous: bat sonar. At the time, it was daunting trying to figure out a way to characterize this sound-based system in mathematical form. But looking back, he says, it was comparatively easy to figure out the waveform bats were using. It turns out to be "optimal for finding yummy moths," he says. "That's a simple problem. It's a sonar problem—you can actually write down the optimal solution—it's a formula." The waveform each kind of bat used was just right for the environment it lived in and for determining the kind of dinner that was likely to be flitting around nearby.

Rather than working with sound waves to elude predators, octopuses are artists in the field of light (apologies to the late Thomas Kinkade). "The cool thing about this," Baraniuk says, is that "it's nothing you could ever write down in an equation." Cool—or if you're not a signal-processing engineer, headache inducing.

Teams of researchers are working to better characterize the oceans' landscapes of luminosity, these so-called light fields, on a mathematical level. But octopuses are a million steps ahead of us and our clumsy computer programs and mathematical models. These animals are somehow automatically processing this chaotic distributed light. "The idea of understanding the light field that's welling down on you from the sun is an extremely hard problem, and the fact that they are able to do it so well is pretty extraordinary," Baraniuk says.

As he talks, he starts to survey our surroundings. Up in the fresh air of an unseasonably warm November afternoon, the sun is shining, and there's not even much of a breeze to ruffle the leaves of the oak trees on campus. Baraniuk concludes: "This is pretty static. If the wind was blowing, it would be a little less static." Even with a gust, it would still be relatively simple up here, because the light is coming, for the most part, from one direction: the sun. Sure, it's being scattered by the atmosphere a little and reflected off various surfaces, but even at night in a room with several different lights on, the light is coming from discrete individual sources. Not so underwater. Underwater, the light problem is even denser.

"Underwater it's a mess," he says. "Underwater, light's coming from all directions, because it's scattered by the water."

You can get a sense of this problem next time you look down into clear, shallow water in bright light. Rippling planes of light dance off the top, the bottom, and inside the water column itself. They're in constant motion with the waves as the changing water surface angle focuses light in different directions. "It's beautiful, right?" Baraniuk says. But to model that mathematically, "That's complicated, man—super, super complicated."

It creates quite a mess for those trying to sort out how the animals make sense of it. But if we can get a better handle on it, Baraniuk says, "from the engineering side of things, this has all kinds of cool applications," such as much more realistic computer graphics and perhaps even better artificial intelligence vision. To get there, though, we're going to need some high-power processing. Assembling a good digital light-field model requires an enormous amount of data. If you take a picture with a camera that's roughly six megapixels, you've captured two dimension's worth of information: two thousand pixels by three thousand pixels, which is 6 million pixels. "If you *really* want to understand the light field, which is understanding the light at a point coming from every direction—and going in every direction—it's actually about seven dimensional," Baraniuk explains. So that's seven thousand-times-a-thousands. "That's like 10 billion pieces of information."

That's a lot of information for one invertebrate brain to process—even if you're an octopus. So to deal with all of that data, an octopus might be using its own skin's layers to help it read the landscape. As

Baraniuk explains, with layers of specialized cells, "you can reject information from certain directions and wavelengths." So an octopus could be calling on a complex web of filters built right into its skin to help sort information before it even comes in for processing. That could help handle "the data deluge problem," Baraniuk notes. Even octopuses, amazing as they are, can't compute the massive quantities of information that would be streaming in from all angles all day and all night.

Smart Nanotech Skins

A growing human awe of—and a mounting quest to imitate—these cephalopods' skills have been bringing some unlikely combinations of scientists together. Being able to perform even half of the octopus's disappearing act could camouflage submarines from sight or sonar, and being able to harness the perceptual abilities of opsinlike compounds could improve practices as common as medical imaging.

Sonars of whales and dolphins have long been of interest to the military, which has studied them to create our own version of this capability—and to develop sonar-evading technologies. But octopuses aren't working with something as simple as sound-based sonar, so they flew under the government's radar, so to speak, until recently. The military is now keen to decipher the octopus's camouflage secrets. The U.S. Office of Naval Research has awarded millions of dollars in multiyear grants to institutions to try to better understand this color-changing code. In the basic biology world, that's a big chunk of change. One group has landed grant money to assemble a sophisticated "holodeck," which can project different backgrounds onto an aquarium's sides to help researchers observe and control changes in the animal's skin in real time. Other researchers in the Defense Department–funded program are investigating the genetics and biochemistry of the proteins responsible for the color change.

"Obviously, cephalopods do what they do for defense—they defend themselves by blending in," Naomi Halas says. So perhaps a similar technology could improve "the safety of the soldiers" by creating a material that's "good—super, super good—at blending," she says, neglecting to mention that octopuses also use their camouflage, at least occasionally, to sneak up on *their* targets.

The researchers at Rice—in collaboration with Hanlon at Woods Hole and others—are working on one of these multiyear, multimillion-dollar research grants from the U.S. Navy. Also working on that project is Stephan Link, who runs a nanomaterials lab at Rice. His task is to help design new building blocks that could sense color and light and then mimic it back. Link explains, with extra German weight on his words, that his group isn't necessarily trying to *re-create* the pigment cells that Hanlon and others are studying. Rather, "we are taking that as inspiration and then trying to see if we can come up with materials that have similar function," he says. To try to accomplish this, Link, Baraniuk, Halas, and their collaborators are scaling down—way down, all the way to the nanometer, which is one billionth of a meter. (Human hairs, for example, are roughly one hundred thousand nanometers wide.)

Within the scale of the nano, some seemingly infinitesimal alterations can make a surprisingly big difference in the way a particle interacts with the world. A change of metal or a slight shift in size or geometry can take a nanomaterial from dud to dynamo. For example, a particular cluster of nanoparticles of a certain size, shape, and number, with a lot of free electrons, will create almost a buzzing electron cloud. These electrons, in Halas's words, "slosh around in response to the light wave." But it won't just be random vibration. It will actually be resonating at a particular frequency, one that can absorb or scatter light in different ways. If they are engineered to have oscillations that cancel each other out, light will pass right through. Or you can make them highly absorbent and use the size of the particles to control what color they take in. "So we can make structures of all different colors that are based on some little metal geometry," Halas says casually, as if she were talking about Legos. Although these constructions are clumsy answers to the octopus's elegance, scientists hope that they will one day be able to mimic the quick responsiveness of the octopus's camouflage to changes in its surroundings.

Halas first got interested in the crazy problem of cephalopod skin (or perhaps more accurately, from an evolutionary perspective, the amazing *solution* Nature has devised in the form of cephalopod skin) at a scientific meeting. It was the first time she encountered Hanlon's now-famous video of the startled octopus jetting away from a perfect plant camouflage. At the meeting, the video was presented as an example of challenging materials researchers should think about tackling in the next ten or twenty years. And that got Halas thinking. "Mother Nature has figured out how to do this—could we?"

So she started to think about the problem (or solution) on a deeper level: "So what does that mean? Does it mean making material that changes color? Sure, that's absolutely a challenge. Although people have done it." But how do you make material change color instantaneously to match its background?

She mulled over this issue until she couldn't stand it anymore. "I literally called Roger up one day, out of the blue, and started talking about this." It turned out that she, a nanophotonics researcher in Texas, and Hanlon, a biologist in Massachusetts, had mutual collaborators in, of all places, the Air Force.*

At the time, Hanlon and his colleagues were just starting work on the issue of cephalopod vision and how the animals might be able to detect light and colors through sensors outside of their eyes. Halas says she was transfixed by the idea of this, as she calls it, "extraocular vision." So, she recalls, she got "this funny idea" that they could develop a material that "sees" color and changes its own color to match—and do it all without having a "brain" or central computer or circuit board to process the information. These active materials would "provide a local, intelligent response that would change the properties of a structure so that its optical properties will change." Sound impossible?

Apparently, it's not. Cephalopods might already be doing this. Of course, evolutionarily, the octopus has had much longer to come up with this ability than we have had to understand it. "Nature has had 3 billion years to do this; we only have a three-year grant period,"

* Squid can fly—a few of them can jet themselves out of the water a few feet—but the Air Force was more interested in cephalopods' skin-camouflaging abilities.

Halas says. "So we work a lot faster, but we don't work as nicely as Mother Nature does."

She hopes that in the end they will be "changing the definition of 'material,' because it's a detector *and* it's a material," she says. "The boundary becomes blurred."

This future network of materials will, of course, not just have to turn brown on a brown background to work as camouflage. Like octopuses, it will have to detect patterning and variation and then replicate it. But in Halas's eyes, this is not necessarily the most challenging part. Instead, "subtle changes in color are much harder to perceive than pattern recognition," she says. "That in itself is a really exciting problem."

Even if in five or ten years biologists discover that much of this lightning-fast color changing *is* controlled, somehow, by the octopus's central nervous system, that's okay. By then, perhaps engineers will already be on their way to developing a material that is based on local control for our own uses. "So the way we connect to biology is not in a strict biomimetic sense," Halas says. It's more in the functional sense.

Pondering the challenge of this do-it-all material, Halas initially thought the sensor would be the biggest challenge. People had, after all, already made color-changing materials. But a nanomaterial that creates a particular color to disappear against a background must first detect the colors around it. By borrowing technology from another project Halas had been working on—one that detects light via nanoantennas and turns that energy directly into electrical signals—they were able to make a big early jump in the detection direction. They published a description of the device in *Science* in 2011. "This is actually *something*—we have an eye!" Halas says, still excited from the development. These tiny nanoscale antennas don't just see light, they also can detect certain colors. Other sensors often need additional components to filter out various hues and waves of light on the color spectrum. However, the researchers still have to get this "eye" to communicate what it is sensing to a material that can actually change color. "This is where it's going to get really, really interesting," she says.

There has yet to be an obvious solution. In fact, this is where the nanoengineers have come skidding right up alongside the biologists who are just now trying to figure out how the octopus is making this connection. This environment of mutual discovery makes for particularly fertile scientific ground. With many of these problems remaining unsolved on both sides, it opens up the fields for all kinds of ideas—some of which might sound outlandish but that might just be crazy enough to work.

To make this project even more challenging, Halas and her colleagues are trying to create this magic material in a form that would work well in our world. Octopuses, squid, and cuttlefish have covered themselves in a soft, chromatophore-filled skin. But as Halas points out, "You couldn't take the same structure and put it on the side of a car."

This nanotechnology engineering might also eventually be translatable to nonvisible wavelengths, such as radar and sonar. As revolutionary as the stealth bomber was in the 1990s, its radar evasion was still passive, relying on its physical shape and the current radar technology to fly largely undetected. But what if a plane—or a submarine or tank—could actively disappear? It would be as if the moths that bats hunt had a sonar system of their own that would absorb the bat's call and return one at just the right time and frequency to let the bat know that it's just empty airspace. In that scenario, "bats are never going to know if there's anything there," Baraniuk explains. "Totally amazing." If a plane or a sub could do that, it would truly vanish off radar or sonar screens.

Borrowing from the octopus's distributed sensor network model, Baraniuk says, one could also someday make wallpaper with artificial sensors distributed in it that could "sense activity in a room, how many people in the room, and what they were doing," he says. Discrete, big brotheresque video cameras might someday seem quaint in comparison. As Baraniuk describes the visual sensing ability of these mechanisms, even in octopuses, he says he hesitates to use the word "dumb." "But they really are—dumb cameras" that don't have fancy lenses or refined focusing abilities. Instead, they could be created en masse and tuned to look for certain cues, including ones outside of our visual spectrum, such as infrared or X-ray.

Such sensors could also help us adapt to changing conditions. There are already transition lenses for eyeglasses that darken when it's sunny and windows that do the same thing. Link, for one, doesn't see it as too huge a jump to move from an optical response to a thermal one. It's just a matter of broadening your scope and thinking, he says. Nanomaterials that follow the same pattern could be integrated into textiles. So a smart-material shirt could reflect light and keep you cool when it's bright and warm outside and absorb light and warmth when the sun ain't shining. Or they could even be tuned to change based on ambient temperature. So why couldn't we knit a nanoenhanced sweater?

Halas and her colleagues are also already dreaming of smart sensors that could provide real-time feedback on health. Halas envisions an opsinesque nanomaterial that could warn doctors or patients when it detected errant blood sugar levels or the slightest sign of heart trouble. A visual signal could let a patient know, gently, that it was time to get something checked out before it progressed to a major problem later. Engineered opsinlike visual sensors could also be enlisted to work like simple internal cameras—*Fantastic Voyage* style. They could even be injected into the body to capture live, fine-grained information, such as the status of a blood vessel lining to warn of hardening or buildup.

More sophisticated sensor networks could also help measure broader changes, such as those in the environment, including air pollution, the health of the oceans, or climate change in general. With an integrated patchwork of localized smart sensors, we could tackle these big analysis problems at the point of collection. Otherwise, the amount of data so many sensors would generate could overwhelm a central processor: "We don't have big enough computers to handle all of the data," Halas says. So to take a lesson from the octopus, a lot of it could be processed locally, instantaneously.

Scientists are perpetually having to walk the line between giddy optimism and funding realism when they try to project a timeline into the future for their projects. Baraniuk says he could see some active camouflaging prototypes emerging by 2016, but, hedging his bets, he also cautions that this sort of technology can move in fits and starts.

As this project has brought together people from the wet world of biology with those from the usually dry territory of engineering, both sides have been learning more about how to approach science itself. "The real exciting things in science," Halas says, are the connections you make "that are not what you wrote in the proposal—it's something you realize later."

One thing Halas has learned from working with octopuses is that they continue to outdo our species' best and brightest. "They are humbling," she notes. "It kind of blows you away, all of the amazing things that they can do. And then you say, well, what can *you* do?"

But as some researchers try to emulate these quick and complex capabilities, others are trying to understand just what the octopus knows.

CHAPTER FIVE

Brain Power

The octopus is widely held to be the smartest invertebrate—underwater or above. It not only passes many tests of intelligence, but it also seems to engage in complex behaviors, from play to navigation to tool use. And perhaps most impressive of all, it does these things on its own, without learning from other octopuses.

Even though the octopus has fewer nerve cells and its brain structures are far simpler than ours, it is thought to be a good deal more intelligent than many creatures with backbones. The octopus has about 300 million neurons, according to current assessments. That collection pales in comparison to a human's estimated 100 billion. But it is an Einsteinian leap from even the frog, which has some 16 million—or the octopus's humble mollusk cousin the pond snail, which has about eleven thousand.

But neuron stats are not the only way—and perhaps not even close to the best way—to measure intelligence. Scientists have put octopuses through their paces, testing their ability to open canisters, solve mazes, and even recognize individual people. Octopuses have not only passed these tests but also continue to surprise even those who have worked with them for years.

That was what biologist Roland Anderson found when he helped to design an experiment for a female giant Pacific octopus (*Enteroctopus dofleini*) named Billye. Anderson worked with octopuses at the Seattle Aquarium for more than three decades. His charge, Billye, was hungry for a snack. But she was not going to get any easy handouts. A fellow female octopus named Pandora had gotten tasty hunks of herring in a screw-top jar, which over several tests took her an average of two minutes to open. Anderson says that after having seen Pandora so adeptly learn how to open a standard bottle, he and his colleagues "decided to push the envelope a little bit and offer them childproof pill bottles." So to get the fishy treat, Billye would have to push and turn the bottle's lid at the same time.

Anderson and his team at the aquarium had drilled small holes in the plastic bottle so Billye would be able to get a whiff of the meat inside. Fueled with this incentive, she managed to open the bottle the first time (without instruction) in less than an hour (about fifty-five minutes). And after that? Child's play. With a little practice, she was able to get it down to an average of about five minutes—better, in some instances, I would imagine, than an adult human. After so many years working with octopuses, Anderson, who retired from the Seattle Aquarium in 2009, says that of all the feats he's witnessed, "that's about the most impressive."

Rather than figuring out *if* octopuses are intelligent, "the trouble lies in actually measuring and quantifying octopus intelligence," Anderson noted in an article about his experiment with bottle-opening Billye.

That octopuses can learn is no longer a surprise to researchers. Along with many vertebrates, as well as bees, an octopus can learn to distinguish vertical and horizontal bars and pick out the difference between the letters *V* and *W* (though no one, to my knowledge, has tested them on their ability to distinguish a VW from a BMW). Although they don't always progress in their studies as far as many vertebrates, some octopuses have been shown to learn more quickly than a bird or a rat. Looking at how *much* an octopus can learn, Anderson places them somewhere in the middle of the birds—maybe not as smart as an African gray parrot but probably more intelligent than a chickadee.

And as octopus keepers have known for quite some time, the animals can even learn to recognize individual humans—especially if one of them usually arrives with food. That person will likely be greeted by an eager octopus waiting at the front of its tank.

Filmmaker Jean Painlevé recalled learning this from an octopus that he worked with while he was an assistant at the Station Biologique de Roscoff, on the Brittany coast in France in 1925. "Because I liked the creature, every day at 11:00 I gave it a chicken egg," he explained in a 1988 interview for a French television series. "It grabbed it and went black with joy. . . . It would take the egg to the back of the tank, break the shell, and suck out the contents. One day I gave it a rotten egg. It grabbed the egg, turned black with joy, and went to the back of the tank." But it quickly realized something was amiss because "then it turned entirely white from fear and anger," he recalled. "It shot the remains of the egg at me, over the side of the tank. From that I concluded that it possessed a certain intelligence—which is anthropomorphism," he added (but followed up quickly with the assertion: "We're allowed and obligated to use anthropomorphism. Otherwise we couldn't appreciate anything around us.").

Jennifer Mather, the cephalopod behavior expert from the University of Lethbridge, experienced an encounter not unlike Painlevé's while she was conducting research in a lab in the South of France, she explains to me one evening on the phone from her home in Alberta, Canada. She had grown up near the ocean in Victoria, British Columbia, and was always fascinated by ocean life. Years after developing an extensive childhood seashell collection and cultivating a growing interest in ocean life, she lucked into an animal behavior course in college. What better complex organism to study than the octopus, she figured. So she embarked on serious cephalopod studies in graduate school in 1969 and has been increasingly fascinated with the animals ever since.

One day in 1982, she was deep in her research in the damp lab in France, perched on a stool to observe the behavior of petite musky octopuses (*Eledone moschata*). Next to her were lidded individual tanks of common octopuses, who, as she recalls, were quite active. "So while I was sitting there watching the *Eledone*, I looked over at this one octopus that was trying to push its way out of its tank," she says. "Without thinking about it, I banged on the top of the lid, and the animal went back into its home." All well and good, right? Not exactly. "The next day I came back to watch the octopuses," she says. But as she sat on the stool by the aforementioned affronted common octopus's tank, eerily "the lid lifted up a little bit. The octopus brought up its funnel to the crack, and it jetted a jet of water at me."

Let that be a lesson: Never cross a common octopus. Or a giant Pacific octopus, for that matter. Anderson recalls a similarly soaked night watchperson at the Seattle Aquarium. In her nocturnal duties, the guard would shine her flashlight around the darkened exhibits. Apparently, the giant Pacific octopus in residence did not appreciate these midnight checks, so it took to squirting water on the guard whenever she would enter.

But can an octopus discern *between* people? To test this, Anderson and Mather worked together to create a study that used, as Anderson describes it, "the good-cop, bad-cop scenario." One person would approach a giant Pacific octopus and harass it with a bristly stick, and the other person would approach and then feed it. After the octopus had been accustomed to these interactions over a couple of weeks, when the "bad cop" entered the room, the octopus "would shrink back into the corner, it would turn its suckers out to be ready to fight, and it would blow jets of water toward that person," Anderson explains on the phone from his home on the West Coast. The octopus would also often put on the eye bar display, signaling irritation and aggression. For the food-bearing good cop, though, the octopus "would come up to the surface or raise arms up toward the surface," ready to receive the food, he recalls.

Memory is one thing. Some octopuses, many people claim, are even psychic. Paul, the common octopus living at the Sea Life Centre in Germany, apparently predicted the winning team of all seven games during the 2010 FIFA World Cup that Germany played in as well as the final match between Spain and the Netherlands. Before a match, handlers would give Paul two clear boxes. Each box contained a yummy mussel and displayed the flag of a country competing. In each instance, he chose to first eat the mussel from the box of the country that would go on to win.

Certainly most octopuses are not as prescient as Paul, but there is little doubt that there is something sentient behind those big, watchful eyes.

Researchers and casual aquarium visitors alike acknowledge the almost uncanny feeling of being observed—perhaps even examined—by a cephalopod on the other side of the glass. It's a strange sensation to be so closely watched by an invertebrate, a creature so removed from us on the evolutionary tree. Even animals that have backbones don't often interact with us on that level. After all, when was the last time you felt studied by a frog?

As Michael Kuba, of the Department of Neuroscience at Jerusalem's Hebrew University, explains, there's a positive feedback we get when we look into the eyes of an octopus. They move their eyes, check us out, look back at us. It's an experience similar to interacting with an attentive dog, as it looks to us for instruction. "You come into your lab in the morning and there are ten octopuses sitting in the front of the glass looking at you," he explains to me via Skype video chat from his home in Israel. "You are so excited that you think, 'Wow! They are so curious. They are so smart.' It's the same way when I look at my dog, and my dog looks back at me, and I think 'Wow! He understands everything,' " he says (adding that more likely, "the dog just really thinks 'When will they feed me again? Will they feed me again?' ").

But we haven't been breeding octopuses to respond to our cues, as we've been doing with dogs. Even cats don't often give us quite as much scrutiny. This engagement is part of the reason, Kuba notes, that octopuses are presumed to be so smart. Some of the things insects or other invertebrates do are just as amazing as many octopus feats, but the fact that an octopus seems to look us in the eye catapults it into a different category of intelligence—at least in our eyes. But *do* they deserve as much credit as they've gotten over recent years?

It depends on how you want to measure smarts. Kuba and his colleagues have shown that a turtle can perform a discrimination task, such as learning visual cues to find food, about three times as quickly as an octopus. "When you tell people that your turtle is smarter than the octopus, everybody laughs at you," he says. But an octopus, he points out, provides researchers with more avenues of study than terrapins or many other less behaviorally—and physically—flexible animals.

We also bring to the table our own biases about *how* these animals think. It's easy, and perhaps only natural, to assign a familiar—that is, human—framework to animal behavior. Is a bee *thinking* back to a previous excursion to find its way back to a pollen-laden blossom? Is a mimic octopus planning ahead of time to gather its arms together behind it, flatten its body, and swim just like a less appealing sole in order to dissuade predators from eating it? Many creatures' actions are probably not the result of conscious thought, planning, and execution as we think of them. Rather, the navigating bee and a sole-mimicking octopus are likely the product of eons of accumulated selective events that have created an ingrained behavior that gives each practitioner a leg—or arm—up in life.

Of course, assessing animal intelligence is an inherently thorny endeavor. Going on looks alone, one might be easily fooled as well. Even though the octopus looks like it's all head, most of that bulbous blob atop its arms is body. The central brain is wrapped around its esophagus, and to mire matters more, most of its "brains" are actually in its arms (but we'll get to that in the next chapter). Equating brain size to intelligence can be a fraught assay anyway. In vertebrates, brain size seems to increase exponentially with body size, so scientists have devised an equation for relative brain capacity—a shorthand notation for expected intelligence—known by the euphonic term encephalization quotient (EQ). Humans have the highest EQ—about 7.4. Dolphins hover at about 5.3, cats at 1.0, and rats at 0.4. The score then descends all the way down to approach zero as the brain gets smaller in comparison to the body. Octopuses don't rank terribly high in the EQ ratings, the common octopus garnering an EQ of just about 0.026. A giant Pacific octopus, which can have an arm span as long as fourteen feet, has a central brain about the size of a walnut or two. That might sound pretty tiny, but as Anderson points out, that brain size is about the same as that of some of the very biggest dinosaurs.

Standard tests come up lacking too. If the debate surrounding human IQ is any indication, defining and measuring intelligence is a matter of ongoing—and occasionally extreme—contention. "IQ tests are notoriously bad for measuring capabilities and actual intelligence of people," Anderson notes. So if we still have not arrived at a satisfactory measure of human intellect, how can we measure that of an animal—of an invertebrate no less? "Of course we have no IQ test for octopus," Anderson says. When you cannot communicate with a test subject—or even really understand how it apprehends its world—figuring out *what* to gauge and *how* to gauge it can lead to fairly rough and conflicting assessments.

Curiously Playful

Head-turning intelligence is turning out to be far less rare in the animal world than we once presumed (which makes *us* the dummies for assuming we were the only ones with any wherewithal at all). Bees can, of course, learn where to find flowers, and even snails can be conditioned to stick out their breathing tubes on command. But not all intelligence is created equal. Social insects, such as honeybees and ants, might seem capable of behaviors that you wouldn't think something with a sesame seed-sized brain would be able to execute. As Jean Boal, a biologist at Millersville University in Pennsylvania, points out, although we're still learning about these insects' intelligence, they exist in what is called a superorganism. As members of such a finely tuned system, much of their behavior is genetically programed to fit into a complex social organization, and we don't really know just how smart the individual is on its own. Because octopuses are decidedly antisocial animals, their intelligence and know-how are self-contained—perhaps even more so than our own.

I meet Boal one sunny, early December afternoon at her first-floor campus office. She started out studying not cephalopods but computer science. Slowly, she says, she developed an interest in "wet intelligence," which is computer engineerspeak for the messy, mysterious processes that happen between our—and other animals'—ears. While working on her PhD at the University of North Carolina, Chapel Hill, she happened into a lecture about octopus suckers, she recounts. "And in their introduction they said, 'As we all know, octopuses are really smart.' And I said, 'What? What?! We all know *what*?'" she says with her easy laugh that is quick to dry up and get back to business. "And lo and behold, I ended up doing my PhD on octopus learning."

Boal landed at the small school just outside of landlocked Lancaster, Pennsylvania, in 1999, in part because she's, in her words, "hopelessly scientific." She had been working at the now-defunct National Resource Center for Cephalopods, run by the University of Texas Medical branch in Galveston, Texas. Soon, however, the pressure was on to bring in big bucks in research grant money. So she started to look around for a move, but Amish country hadn't exactly been on her radar. "I'd never heard of Millersville, but they were looking for somebody half marine biology, half animal behavior," she recalls. And here she is still, thirteen years later.

During her many years investigating animal intelligence, she has learned that it's an inherently tough intellectual exercise—for *us*. "We're very myopic," Boal says. "We should be thinking more broadly" about intelligence. But when I ask her how we should go about getting to the bottom of what an octopus "knows," or at least is capable of doing, without imposing our own boundaries of intelligence types on them, she pauses in silence and sits very still, as if pondering how to answer a question that she herself has been fretting about for more than a decade. That, she notes, is difficult.

We are prone to anthropomorphizing our eight-limbed subjects. But the creatures seem to be unquestionably curious. In fact, Aristotle's famous quip about their stupidity might simply have been a misunderstanding of the nature of their intelligence, a common human error of seeing the world only as it relates to us, rather than in its own context. So it might have been reasonable to assume, as he did, that "[t]he octopus is a stupid creature, for it will approach a man's hand if it be lowered in the water." In Aristotle's world, an octopus was there primarily to be caught for food; thus, the animal *should* be concerned with avoiding contact with humans. But rather than intentionally throwing itself into the outstretched hand of death, this ancient Greek octopus was doing what most octopuses do—and what we humans do too—exploring its world. It was probably just curious about this five-fingered piece of flesh that suddenly appeared in its watery realm.

Aquarium visitors see this behavior on display all the time. Given a new object, such as a bottle or jar, an octopus will examine it with its arms, suckers, and mouth. Countless online videos show them approaching objects and even people not just for close visual inspection, but also for full-on, physical investigation.

One underwater video documents an octopus absconding with a diver's still-rolling digital video camera. Aside from some moments of darkness—during which the lens was presumably covered by the fleshy web of the thieving octopus—the footage reveals a probing landscape of suckers as the octopus fondles the foreign object. Taking advantage of the animal's apparent curiosity, the diver was able to lure the octopus's attention safely over to the shaft of his spear gun and snag his camera back. (Getting the curious creature off of the spear gun proved to be another task altogether, so instead, the driver took the octopus for a little ride while he swam around.)*

Tamsen DeWitt, a biologist at the Invertebrates Exhibit at the Smithsonian National Zoo in Washington, D.C., has entertained her many octopus charges over the years with plenty of oddball objects, including dog toys, a Mr. Potato Head, and hamster balls (sans hamsters, of course—a little crabmeat inside makes for a better treat). Each octopus has a different favorite. One octopus, named Caroline, loved a red Kong dog toy. DeWitt, who goes by Tamie, has been working at the National Zoo for more than twenty years but is as energetic as if it were her first

* All of this action to the soundtrack of Dalmatian Rex and the Eigentones' "Octopus, I Love You."

day on the job. "I've always been with invertebrates—they are my *favorite*," she says emphatically.

"We are really trying to find out what is the most stimulating object for the animal—what keeps them most interested," she says, when I talk to her over the phone before I visit the zoo. "They are so intelligent, we want them to stay stimulated so they don't get bored. With some mammals, enrichment is such a big topic right now in zoos." When I visit the National Zoo, for example, the elephants are getting an entirely new and improved enclosure. But it's taken longer to realize how important this is for less familiar—or obvious—animals.

After so many years of working with the octopuses at the Invertebrate Exhibit, DeWitt says she is still frequently surprised at what tickles octopuses' fancy. One object she thinks will be dynamite is dismissed as totally uninteresting, whereas something else she thinks might be dull will still be of keen interest to the animal the next morning, and someone will have to drop food in the tank to distract the octopus long enough to retrieve the object.

This playful curiosity can make research a bit difficult, however. Simply setting up tanks and equipment that octopuses won't slink out of or break is a serious consideration for any octopus experiment. Having backup items on hand is a must, because one of the subjects will inevitably pull something into the tank that it shouldn't, notes Boal. "You have to be prepared for that," she says, so that you can pick up another tool to use on the next octopus until the first one loses interest and you can recover your original piece of equipment.

Some octopuses will tire of a new object quickly, but others will toss it around for quite a while. One giant Pacific octopus, described by Jennifer Mather and Roland Anderson, discovered that it could blow a plastic pill bottle with its funnel toward the tank's water input stream, which would send the bottle back to her. Anderson called Mather right away to describe the scene: "She was at one end of the aquarium, and the water inflow was at the other," Mather says. "The water inflow pushed the bottle toward her, and she pushed it back, and it came back to her. She did this about eighteen times. 'She's bouncing the ball!' "

Anderson told Mather excitedly at the time. Despite not even being there to see it in person, Mather describes this as "one of my favorite minutes in all my research career." And with a career that spans more than four decades and countless octopus subjects, that's no small statement.

The octopus did this periodically for days, seeming for all intents and purposes to be entertaining herself. And as Mather points out, if she had simply been trying to rid herself of the pill bottle's presence, she likely would have taken a different approach. For instance, a male octopus in the same experiment, apparently finding the bottle displeasing, held it at the tip of its outstretched arm "as far away as possible," Mather says.

Another octopus researcher, James Wood, who runs the Web site The Cephalopod Page, once had an octopus that managed to amuse itself without any such toys at all. Wood is a biologist from south Florida who says he was born a marine biologist. By age three, he says, he was raising tadpoles, and by his early teens he had moved on to keeping locally caught marine fish in his bedroom aquariums. Most kids in the area went fishing and surfing—and Wood joined in, but he describes himself as "kind of a geeky surfer." When the waves weren't very good, he would poke around in the seaweed and turn over rocks to study the ocean life.

Soon he was keeping octopuses. The first octopus he ever caught had an impressive way of entertaining itself in the aquarium. It would spread out its web and suction itself onto the bubbler at the bottom of the tank. Once its web was full of air bubbles it would let go and shoot to the top of the tank. Wood watched it do this again and again. Was it playing? "I mean, it sure looked fun to me," he says, noting that if it had been a dog, we would surely have called that behavior play.

How do we know this is "play"? We don't, of course. But some researchers are convinced that play in animals is in fact much more common than we think it is. The reason that we don't see play in more animals, Mather thinks, is because we're looking for play as *we* know it. "So, for instance, we don't know if we've ever seen a lobster play," she says. "I don't know what a lobster would play with *if* it played.

We were just lucky because the sort of thing the octopus did was easily recognizable." (Although some researchers, including Michael Kuba, note that there have actually been fierce debates between the octopus researchers and lobster researchers as to who has the most intelligent subject. When I expressed my disbelief, he said that some lobster devotees who are terribly in love with their lobsters—"of course not only in a culinary sense," he says—have found them to be "quite amazingly smart animals. They just hide it well.")

In the wild, play could actually be pretty risky—it takes energy and also means potential exposure to predators. That's likely one of the reasons, Kuba says, we have seen animals "playing" more when they're in the safety of captivity than in the wild (we also just watch them more closely when they're in our charge).

We and other animals, such as dogs, engage in social play as a way to test out and learn about social rules and relations. But because octopuses are pretty much antisocial, "social play is automatically off the menu," Kuba says. The other two types of play—locomotor play, such as a young horse galloping around a meadow, and object play, such as a cat playing with a dead mouse—are more like what octopuses might be doing. These activities could help them prepare for the future and maintain top neurological function, just as we do a crossword to keep our brains going or toss a ball around to keep our body in practice. Play should be something that is pleasurable—not just that it might make you smile but, more important, that it has made you more fit to face challenges in the future. So if an octopus is in a tank where "things are absolutely dull and boring," play behavior might well be just a way for the animal to keep its complex brain and nervous system busy, Kuba says. "Life is not built for stagnation."

And as some scientists and pet owners have noticed over the years, an octopus doesn't seem to do so well when it *doesn't* have something to play around with (and perhaps hasn't figured out the bubble ride as Wood's octopus did). When Boal first started her work on octopuses, she says, her subjects weren't being very cooperative. Many people suggested that the octopuses weren't performing well because they were stressed by too much stimulation in the bustling lab. They advised that she "needed to calm down their environments," she recalls.

But one of Boal's undergraduate students, Marie Beigel, noticed that lab octopuses kept in individual, somewhat spartan tanks didn't look so great. And after learning more about octopuses' smarts, she wondered, in a 2006 paper she and Boal published on the topic, "if octopuses are so intelligent, how can we limit these complex invertebrates to a bare and confining tank?"

So Beigel and Boal set up an experiment to figure out whether the octopuses' dull living quarters really were getting them down, or whether, as earlier researchers had told Boal, it was the busy lab that was stressing them out.

They constructed two types of tanks for six California mudflat octopuses to try out—one tank that was considered "impoverished" and another that was more "enriched." The impoverished environment was a pretty standard lab setup and was by no means monastic. These small octopuses were each in an individual clear plexiglass tank (about one and a half feet by eight inches by about a foot), which was outfitted with a large terra-cotta flowerpot (for a den), a smaller terra-cotta flowerpot, one to three small stones, two to four glass beads, and three to five shells. These tanks were situated in the lab so that the octopuses could watch the goings-on around them—and one another—as before.

But for the enriched environment, they created an even more interesting space. Tanks were twice as large, and in addition to the above contents, they had a crushed-coral bottom, plants, and a view of a live wrasse fish swimming in a jar nearby. The six experimental octopods spent two weeks in the standard environment, two weeks in the enriched environment, and then a final week back in the standard environment for comparison.

When the octopuses were observed living in the plainer lodgings, they seemed a little stressed. But hang on, you say. How can you tell when an octopus is on edge? Researchers have developed a list of behaviors to watch for, and many of these octopuses betrayed a range of them, including throwing themselves against the sides of their tanks, inking, flashing angry colors, or turning pale white and hiding for much of the day, just to name a few unpleasant activities. Indeed, in the impoverished environment, the octopuses jetted themselves into the tank walls about three times as frequently as when they were in the enriched tanks. They also seemed to show more stressed-out colors, were less effective at camouflaging, and were less inclined to handle objects than when in the enriched tanks.

The octopuses' behavior was only recorded for five-minute chunks a few times a day. But at other times that were not part of the formal observation, students saw the octopuses in the standard tanks exhibiting even more extreme behaviors, including autophagy—eating their own appendages. Autophagy was never noticed in the enriched tanks, and it seemed to be the worst once the octopuses had experienced the enriched environment and were put back in the impoverished tanks. After all, if you went from a small, spare cubicle to a well-appointed office and back again, you might at least feel like biting your fingernails.

Boal and her students weren't the first to report captive octopuses eating their own arms (which is different from a rodent's chewing off an injured tail or leg, a behavior that likely evolved to avoid infection). This is one of the reasons biologists like DeWitt and others at public aquariums are keen to keep this disturbing boredom-driven behavior at bay. (As you might imagine, an octopus eating its own arms is not so good for visiting school groups.)

Just about everyone who has worked with octopuses for any period of time—or even just entered the room where one is hanging out in a tank—can tell you that they notice when something in their surroundings changes. "Certainly in the lab they're very interested in watching what's going on, particularly the longer they've been in the labs—they get less stressed," Boal says. Overall, in fact, she says, "I think we've been slow to recognize the importance of sensory deprivation on animals."

Handy Collections

Curiosity and apparent playfulness, however, can only get an animal so far. A borrowed underwater camera or fondled pill bottle might be little more than the cephalopod equivalent of a cat sniffing a new piece of furniture or an ant testing an unfamiliar surface with its antennae. Some clever octopuses have been observed putting objects to use as décor—as well as, some scientists argue, bona fide tools.

Like some species of birds, octopuses occasionally decorate the area around their dens with found objects, both natural and man-made. This creative behavior is, of course, the inspiration for the Beatles' "Octopus's Garden," from their 1969 album *Abbey Road*. Ringo Starr wrote the lyrics while on vacation in Sardinia, lounging on Peter Sellers's yacht, as he recalled in the book *The Beatles Anthology*:

> I stayed out on deck with [the captain] and we talked about octopuses. He told me that they hang out in their caves and they go around the seabed finding shiny stones and tin cans and bottles to put in front of their cave like a garden. I thought this was fabulous, because at the time I just wanted to be under the sea too.

This unusual collectors' tendency has not only inspired songs, but it has also helped lead researchers to some amazing discoveries—of the archeological variety in addition to the behavioral kind. One high-tech, $20,000-a-day expedition set out to find ancient shipwrecks off the Macedonian coast of Greece. The research team launched with a well-equipped ship, a manned submarine, and a remotely operated vehicle (or ROV for short, which is piloted from a distance). Despite all of that expensive equipment, the expedition owed some of its big finds instead to a local octopus that had taken up residence in an old clay pot. This hoarding cephalopod had gathered bits of ancient pottery and other odds and ends outside its den. One of the treasures it had brought home turned out to be a rare sauroter, a bronze point from the back of an ancient Greek spear. "Very often the first clue that a shipwreck is nearby is a pile of artifacts collected by these wonderful creatures with an antiquarian's passion for old things," noted one of the researchers in a 2004 article describing the expedition. And fortunately, unlike the giant treasure-guarding octopus in Super Mario's early 1980s Game & Watch video game Octopus, this sunken-goods collector relinquished its prizes relatively peacefully. The Nintendo octopus was inclined to kill its shipwreck-raiding adversaries.

Other octopuses practice more pragmatic curation. Some veined octopuses (*Amphioctopus marginatus*), for example, gather coconut shell halves to use as shelter. This behavior, Mark Norman of the Museum Victoria and his collaborators suggest, is an example of tool use—a feat that would set it far apart from other invertebrates and place it in a select group of only a handful of other animals.

A video, published online in tandem with the 2009 *Current Biology* paper describing this behavior, revealed several of these octopuses off the coast of Indonesia, where the sea bottom didn't offer many rocks for natural dens. One video clip showed a small octopus awkwardly clutching an oversized coconut shell beneath its web while attempting to skitter along the seafloor with its outstretched arms. Another shy subject had two halves on hand when it was startled by the researchers. Noticing them, it quickly snapped the shells together, hiding itself inside.

Although hermit crabs find new shells to live in and other octopuses have been found making use of man-made objects such as bottles to use for homes, these Indonesian veined octopuses, Norman and his coauthors argued, are demonstrating something far more advanced—planning—with a purposeful manipulation of an object to realize a goal: a shelter on the go. That, they say, constitutes tool use.

Jennifer Mather contends that we've seen plenty of other examples of tool use in octopuses, but we just hadn't recognized it as such. The coconut shell is "a really spectacular example, because it really does suggest foresight," she says. "The octopus must somehow know that he is going to need shelter, and therefore it thinks coconut shells could be of good use. In terms of cognition that's *pretty* good."

Not all octopuses, of course, have access to handy, home-sized coconut shells on their home turf. Others seem to make use of more common objects, such as rocks. Some octopuses have been observed piling small stones in front of their den opening—presumably as protection from predators.

Observing underwater octopus behavior in the 1980s, Mather watched an octopus clean its den, then collect rocks to narrow the opening. As she describes in the book that she co-wrote with Anderson and Wood:

> To do this, it must have some idea of what it wanted—known in some way that a pile of rocks would make the den entrance smaller—and then looked out across the sand to see suitable rock candidates and gone out and picked up the right number.
>
> In describing what the octopus had done, no matter how I tried, I found myself needing to include words like *wanted, planned, evaluated, chose,* and *constructed*—words that animal behaviorists of the time (and even now) were not likely to use regarding invertebrates. The words moved the animal out of the category of reactive plodder to that of thinking and anticipating being.

"When it comes right down to it," Mather tells me of this anecdote, "that's tool use." As with play, we've been a bit dense about this sort of advanced cognition in species other than ourselves, let alone those without backbones.

Other researchers argue that there are plenty of behaviors that don't necessarily rely on tool use but that do require plenty of planning—a pretty advanced intellectual exercise. Woods Hole's Roger Hanlon presents some of his astounding underwater camouflage videos as evidence of this sort of planning ability. He shows the video of the octopus in the "flamboyant pose," with some of its arms extended up to mimic plants while it crawls slowly along the bottom with its other arms, looking quite magnificently rocklike. The octopus pauses near one rock and then moves farther away to another one. "This is the kind of thing I would call cognition, which means that the animal is really looking at the whole scene, deciding where to go," he says.

Whether or not they were aware of octopuses' apparent ability to plan, Spider-Man comic creators Stan Lee and Steve Ditko imbued their 1960s supervillain Doctor Octopus with such a notorious proclivity for plotting that some of his aliases included Master Programmer and Master Planner.

Big Personalities

While many researchers have been looking for generalized intelligence among octopuses, other researchers take the octopus's apparent individual personalities as an indication of well-encephalized sophistication.

Back in the 1980s, Roland Anderson had been working at the Seattle Aquarium for several years when he realized that keepers and volunteers were bestowing names upon only a few types of resident animals: the seals, the sea otters, and, oddly, the octopuses. This, he suggested in a 1987 paper, was indicative of the animals having individual behavior patterns distinct enough that we humans could pick up on them. The Seattle Aquarium has since been home to some famously—and infamously—named octopuses over the years. The notoriously reclusive Emily Dickenson was eventually relieved of her public life because she refused to emerge from her tight hiding spot behind the tank's backdrop ("I believe all of her poetry was published after she died," Anderson says with a laugh). Leisure Suit Larry was an overly friendly fellow, who, as Anderson recalls, if he had been a human "would have been arrested for harassment—because his arms were all over you. He was kind of slimy." And then there was the real troublemaker, Lucretia McEvil.

Tamie DeWitt has noticed distinct personality differences in her many charges in her two decades of working with octopuses at the National Zoo. Their previous octopus, Octavius, came with a reputation that preceded him. The diver who collected him near Vancouver warned her that "this octopus is really spicy—he tried to climb out of the bag!" she recalls. And he kept up his antics once he arrived at the zoo. In order to acclimate a new arrival, aquarium workers add water from the tank into the octopus's bag and then put the whole bag into the tank. Some animals seem to just sit around for a while getting used to things, but this one "just shot out—and they never do that," DeWitt says. The zoo's octopus tank has a hiding area in case the inhabitant wants to get out of the public eye for a while. But that octopus never used it: "It was always out—as soon as you open up the tank it comes right up." Not all of their octopuses have been so outgoing, however. One, named Shadow, vanished into the backdrop of the tank and went into extended hiding. The aquarium workers hadn't even noticed the opening it found when they had set up the tank. "It couldn't have been the size of a dime," DeWitt says of the hole. But sure enough, this shy octopus found it and made good use of it.

Just how does one go about ascertaining an invertebrate's personality? In the absence of a Myers-Briggs test for cephalopods, one early personality test looked for nineteen different behaviors of wild-caught East Pacific red octopuses (*O. rubescens*). Anderson and Mather observed them when they were put on alert (by opening the tank lid), fed (treats dropped in the tank), or threatened (by touching them with a brush). They found that individual octopuses did, in fact, react to each situation very differently. When threatened with the offending brush, for example, some touchy octopuses would squirt ink at it and jet away; other more daring ones would attack it. With this, the researchers established three aspects of octopus personality: boldness, shyness, and passiveness.

Another study sought to drill down to see just how much of these individual differences might be genetic rather than shaped by life experiences—a debate that is still heated about humans. While he was a master's student at Portland State University, David Sinn tested personality traits in *Octopus bimaculoides* that were born and raised in isolation in identical bare tanks. He and his colleagues found that even these octopuses seemed to have four different personality dimensions (engagement, readiness, aggression, and avoidance) despite having essentially the same upbringing. This suggests that individual behavioral traits in the octopus, like those in humans, are at least in part genetic. In the octopus's case, differences in one brood might be partially explained by their having different fathers.

For his part, Hanlon is not buying into this cult of personality. He prefers the term "behavioral syndrome." He doesn't take issue with Sinn's findings showing stereotypical *behavioral* differences in individual octopuses. "I just don't want people going out there and start using 'personality' and comparing them to mammals and primates," he says.

Jean Boal generally agrees with Hanlon on that point. She discovered these individual differences, whatever you might prefer to call them, while she was still in graduate school. The octopuses in the lab where she was working had rather discerning palates and wouldn't eat their frozen squid food unless it was relatively fresh. So Boal herself quickly learned to tell fresh from unfresh frozen squid (presumably without tasting it). But unfortunately, the supply of high-grade squid wasn't consistent. One day she knew she was bringing her subjects subpar chum. She approached the first tank in the row, where a large female lived, and dropped some of the food in for her. She then proceeded down the line of the tanks, feeding each octopus. When she was done, she walked back to check on the large female in the first tank to see if she'd eaten the squid. "She was sitting at the front, watching and waiting," Boal says. "And when I came, she looked at me, and then she swam over to the drain for the tank—watching me like this," Boal widened her eyes as she acted out a slow, deliberate sideways walk across the room, "and pushed the food into the drain." Boal got the message loud and clear.

Perhaps the personalities are only a suggestion of something deeper within the octopus's intelligence that we are drawn in by—a developed consciousness perhaps, or at least a familiar curiosity? Maybe the very richness of behavioral traits combined with those engaging eyes and dexterous limbs are some of the reasons we are so much more engaged by octopuses than by horseshoe crabs or lobsters. Biologists working with these animals tend to bond to them differently than they do to, say, a resident starfish—even if that starfish actually sticks around for much longer. After just one year at the Poulsbo Marine Science Center Aquarium in Washington State, Mr. Bob the octopus was showing signs of his rapidly advancing age. "I don't want him to die in a tank," aquarium director Patrick Mus told reporters in 2010 as he prepared to release the old octopus into the nearby Liberty Bay. But before Mr. Bob floated out to freedom, Mus was seen clutching the cephalopod's soft body and giving him a big wet kiss on the mantle—an embrace worthy of a romantic train platform departure.

Quick Learners

Several bird species have proved to be adept learners. A handful of animals, including whales and great apes, are even able to pass along complex skills and information to the next generation, creating a chain of accumulated knowledge. It seems obvious to *us* to learn how to do something by watching someone else do it, but most animals haven't quite got this trick figured out.

Octopuses are a curious case. Unlike most other tool users we know of, they are not social creatures. A baby octopus will never really meet its parents, and each generation grows up and spends their adult lives essentially alone.

A controversial study, published in 1992, reported that octopuses that had observed fellow octosubjects extract snacks from certain colored containers were quicker to learn where to find the food than octopuses that had to figure it out on their own. That would be an example of observational learning—which would be pretty impressive for an otherwise misanthropic mollusk.

Other researchers have since contested the findings, suggesting that the humans who were running the experiment might have unintentionally given the octopuses clues or accidentally left sensory tracks to follow, leading the second group of subjects to the target containers. In comparative biology, to formally qualify as "observational learning," an animal has to be "learning from visual observation of the same species," Roger Hanlon notes. But he doesn't think this means that octopuses don't learn by watching.

"I definitely think that octopuses have the ability to learn by observing," he says. "But that's not 'observational learning,'" because they are, instead, learning things by watching other species that they interact with over the course of their lives. When hunting in the wild, octopuses often use a stop-and-go approach, walking along for a while and then pausing to check out their surroundings before making their next move. "During that time, they're observing, and, I think, they're learning," he says. An octopus might spend some time watching crabs—one of its favorite meals. And by getting a better sense of the crabs' movements, it's possible that "it learned to better capture that crab."

Research on octopus brains has shown that they seem to have a pretty good working memory. "They have short- and long-term memory," Hanlon says. "We use our short-term memory to remember a phone number and e-mail address for a while. They're using it for different stuff."

At the Hebrew University in Jerusalem, Binyamin Hochner and his colleagues are trying to pin down some of the neurological connections that make an octopus brain able to learn and remember—and produce such fascinating behavior. This might help us understand other animals'—and even some of our own—neurology. "I would like to understand how the nervous system controls behavior," says Hochner, who trained as a neurophysiologist, when I talk with him from his Jerusalem office.

"I go first to the octopus rather than to the human cortex." The former is obviously easier to gain access to for live experiments. And because the octopus brain is a tad simpler than our own, "we have a better chance to connect nervous properties to behavior," Hochner says. "This is not an easy task, for sure," he notes. But after decades of research, he says, they are zeroing in on a clearer picture of how information is received, transmitted, and modified during learning and memory. Like many vertebrates, octopuses seem to use long-term synaptic potentiation (LTP), which enhances signaling between neurons, for learning and memory. Hochner and his colleagues showed that when that LTP was blocked or synaptic connections were physically cut, octopuses had a much more difficult time remembering a task they had to learn after the treatment. "But we don't even have the slightest clue to how all of this is going to be combined to control behavior," Hochner says.

In looking for the biological basis for their learning, memory, and behavioral abilities, Hochner and his team have made some of the major modern descriptions of the octopus brain and nervous system. And that has convinced them that the octopus brain is an uncommon one in the animal kingdom. "There are probably not that many brains that are organized to acquire lots of memories," Hochner says. The organization of the part of the octopus brain that controls memory acquisition (the vertical lobe) is reminiscent of that of the mushroom body in an insect's brain, which is thought to also be involved in learning and memory.

The capability for learning seems to be lodged in an octopus's vertical lobe, which alone has some 25 million nerve cells. Better characterizing this part of the octopus brain has been "a step in understanding how the brain is functioning" and how neural networks are organized to carry out higher operations such as learning and memory, Hochner says. And scientists have found that, like humans, octopuses tend to rely on serial memory systems that transform short-term memories into long-term ones.

Now Hochner, Michael Kuba, and others would like to move on from the macro structures in the brain to the transmitters and synapses between nerve cells to understand how these connections make the octopus's vertical lobe so good at learning and memory. They are just starting to find that regulators in the brain, such as serotonin, help stimulate activity in the vertical lobe—a process that seems adapted for sending reward signals to cue specific kinds of learning and memory. But this sort of small-scale work can be tricky with an octopus.

For decades, research on the octopus brain was basically limited to the crude procedure of cutting it up and seeing how that changed behavior. Meanwhile, scientists were learning oodles about the noodles of other animals by implanting electrodes and attaching EEG devices. But with the octopus, "it's very hard to implant electrodes," thanks to those meddling arms "that are very good at removing any foreign objects," Kuba notes. "In octopuses, this is really in its absolute infancy." Kuba confesses to starting the research in naïveté, hoping to jump right into his postdoc recording of the nerve activity of octopuses while they completed learning and memory tasks. "It was a completely miserable failure, because it turns out they just pull the wire." But with improving—and shrinking—technology, he says, we are getting closer. Hochner's research group did manage to implant electrodes in an octopus to measure motor coordination. Nevertheless, "it's probably going to be some time before we get anything like the MRI of an octopus," Kuba says with a laugh.

In the meantime, we can continue to study the octopus in action—and the impressive behavior at which it excels—in order to get a glimpse into how it might be learning. Scientists have long observed various species of octopuses meandering far from home to find food and then return directly to their den for shelter. A couple of Hawaiian octopuses (*Octopus cyanea*) have been spotted wandering way out of eyesight from their home bases, and then jetting straight back home. Giant Pacific octopuses have been observed avoiding hunting grounds they have recently combed over. This sort of behavior suggests that like sharks, primates (some of us better than others), and plenty of other vertebrates, octopuses have decent spatial memory.

But field observation is one thing. Arriving at a defendable scientific conclusion requires testing. And with octopuses, that requires a whole tricky set of special considerations when it comes to lab setups, as behavioral researchers such as Jean Boal know well. A maze that might work for a rat isn't going to work with an octopus. For an octopus, a narrow spot feels like a safe nook, so it probably won't want to budge.

"If you put them in a small space, they say, 'oh, fine,' and they sit down and don't do anything," Boal says. "So you have to figure out what's aversive enough that they want to get out of it, but not so aversive that they will be physiologically stressed."

Some of Boal's main work is on spatial learning, the equivalent of how we might remember how to walk to the nearest market and back home again. So the center of her longtime lab at Millersville was filled with testing tanks and mazes that could hold the cuttlefish or octopuses for various experiments. Cuttlefish resided in tanks on one side of the lab and octopuses enjoyed their individual "octo-condos" on the other side.

To test the octopuses, Boal and her fellow researchers settled on a design called an open-field maze, "which to the average person isn't a maze," she says. It's a big open circle that has small holes in the bottom that can be opened or closed. The (human) team could raise and lower the water level or shine bright lights overhead, making it more or less comfortable for the octopus to be out and about. Because octopuses prefer the safety of a small, dark space, they're motivated to locate an open hole and disappear into it ASAP. Depending on the test, researchers could also change landmarks and the location of open holes in the maze and then film octopuses as they learned to go to different places.

Using California mudflat octopuses for their first experiment in the series of tests, Boal, Hanlon, and their colleagues back at the University of Texas Medical Branch at Galveston set up these circular arenas with one open hiding area to serve as a den. Each octopus was videotaped for three days straight just to see how it explored the area (this allowed the researchers to study the footage closely—and to leave the room so their presence wouldn't distract the octopus or betray any unintentional hints).

The octopuses found the open den relatively quickly, but they also spent a lot of their initial time in the tank checking out their new surroundings. In fact, on average, nine of the first twelve hours the octosubjects spent exploring. But as the hours and days wore on, and the octopuses presumably became familiar with their big, new circular enclosure, they spent less and less time outside of the den. And in the last twelve hours of the three-day stay, they spent only about an hour out. Over the course of the first assay, most excursions out of the den were about an hour long. These sorties were unlikely just for exercise, the researchers speculated (because they became less frequent over time) or for food scavenging (as the octopuses were not fed during the experiment and were likely getting hungrier and hungrier, so any hunting outings would likely have increased with time). "The continuing activity after the initial burst of movement could be interpreted as 'patrolling,'" the team suggested in their paper, published in the *Journal of Comparative Psychology,* to check out the new area, as we might check out a new neighborhood we just moved into.

But to see just how much information about their environment the octopuses were retaining, the researchers put the animals' exploration abilities to the test in another experiment.

The team next gave different octopuses a chance in a similar round-arena setup. The researchers added bricks, small rocks, rubber bungs, a fake plant, and pieces of plastic to the tank to serve as landmarks. Each octopus had twenty-three hours to explore the arena on its own, in comfortable water conditions, with just one burrow open for its use.

After that expedition, the octopus got plopped back into its home tank for a day. Following this day of rest, it was returned to the round arena, which now had a low water level and a bright halogen lamp shining overhead. In this environment, the crepuscular California mudflat octopus was especially eager to find a safe, dark, watery shelter. Twelve of the octopuses got the same burrow open as in the test run. For another twelve, the open burrow was in the opposite location (right side versus left side or vice versa). Six octopuses got both burrows open.

Two thirds of the octopuses first went to check out their original burrow well within fifteen minutes of entering the tank, suggesting that they remembered where it was. (When tested in the same way, wild-caught octopuses performed better than octopuses that had spent more of their lives in the lab.) The results showed that "within one day they can learn the location of an environmental feature that is not needed at the time of exploration but that could have value at another time," the researchers noted. This would be like our taking note of a 7-Eleven when we arrive in a new town—even if we don't feel like getting a Slurpee at the moment.

A third experiment in the paper really put the cephs' spatial memory to the test. A new group of octopuses each got twenty minutes every morning and evening to explore a circular arena with six closable burrows built into the bottom. Each opening was partially covered with a terra-cotta pot saucer so that the octopus couldn't see if the hole was actually open until it got right up to it. These arenas were also shallow and brightly lit, encouraging the octopuses to seek shelter. Each octopus was presented with one consistently open burrow—the rest were blocked, but the octopus couldn't see from a distance whether they were open or closed. And the octopuses were keen to get out of the bright light. "Once the octopus found the open burrow, it would enter and pull the saucer over its head to block the light," the researchers explained in their paper.

The octopuses then got one week off to relax back in their own tanks. But after that week was up, they were placed back in the center of the circular tank to see if they could remember—after the intervening seven days—where to find their familiar open burrow. Impressively, most of them located it with little trouble, and three of them even bolted directly there, which suggests that they had retained an impressive spatial map of the tank for the full week.

For a final test, the researchers closed the old opening and instead made available a burrow in a precisely opposite position. "Most of the octopuses [had] learned the location of the open hole and were disrupted by the reversal," the researchers wrote. But just as octopuses had gotten quicker in finding their initial burrows, they also learned how to locate their new burrow location even more rapidly, showing not only that they passed the maze test but also that they got better at learning with time and experience. Pretty good for a relative of the sea slug.

Although the octopus's spatial learning has turned out to be relatively quick, the progress of the research discovering it has been painfully slow. As Boal recounts, their first maze was developed in the mid-1990s, but their major paper on spatial learning didn't come out until 2007. And this paper was in part just a way of showing scientifically that this sort of test would work for octopuses, and as they noted in the paper, could "offer a step forward in the objective assessment of learning in cephalopods."

Intelligence tests can be slow going not only due to all the painstaking efforts to make them scientifically sound, but also because of the octopuses themselves. Octopuses are, well, temperamental. "Some days they work, and some days they don't," Boal says of her subjects. It can be that "they're just not hungry today, and they're just . . ." she pauses, as if to collect herself, "It just—it takes a lot of patience to work with them and to collect enough data," she says. This exasperation even manifests itself in the formal research literature. As Boal and her coauthors conclude in their multipart maze paper—in proper academic elocution—"a chief roadblock in investigations of octopus learning abilities has been their relative intractability as experimental subjects."

That frustration has driven some researchers out of the lab and into the field, where they can collect anecdotes of octopus behavior in their natural habitats, doing what they do—instead of trying to cajole them into doing what *we* want them to do. These observations add the element of realism lacking in a contrived lab experiment. But it is hard to generate controls for observational work, which makes proffering any substantial scientific conclusions difficult. The two types of research can work together, Boal says. "Observations are great for generating hypotheses that you have to turn around and test."

Some elements of octopus behavior seem so strange and deeply evolved as to be untestable in a laboratory tank. For example, why does a lone veined octopus start collecting coconut shells for shelter, or how does an isolated mimic octopus know to impersonate a peacock sole to elude a predator?

And the impressive cephalopod intelligence isn't limited to octopuses. Cuttlefish have shown themselves to be willing maze solvers, and squid are thought to be quite bright as well. Working with various types of cephalopods, Boal says, has made her realize that there's no reason to think that any one of these impressive cognitive abilities is limited to just one type of animal.

Unexpected Evolution of a Complex Brain

Cephalopods have had hundreds of millions of years to cultivate their intelligence, but the development of exceptional (okay, at least pretty darned impressive) intelligence is not an evolutionary given. Much of the animal kingdom does quite well for itself—cockroaches, mites, fruit flies—apparently without much complex cognition at all.

As Boal points out, "big brains are really expensive" energywise. We humans, for example, spend some 20 percent of our energy just fueling our big brains. So the octopus too must be getting an outsized benefit from its investment. This leads researchers like Boal to the next logical question: "What is it doing that's so valuable?" Obviously, this big brain is not just there to vex behavioral researchers and graduate students. So what's been keeping it so busy over these millions of years throughout so many brief generations?

"It's one of the things we've not figured out: Why an animal with this relatively short life needs to learn so much," Mather says. Perhaps it's the ever-shifting surroundings in an octopus's world that require such an agile mind. "I think it's because they have such complex environments—such quickly changing environments," she says.

Although, as she also points out, plenty of other animals live in the same or similar environments and don't seem to be even on the verge of tool use. Take the humble horseshoe crab, for example. Marvel all you want at the clever cephalopods, but "then you look at the horseshoe crab and you realize that the same model horseshoe crab has been there for millennia and millennia and millennia," Mather says. "They haven't come very far, they don't do very much—and they're doing just fine."

Some researchers suggest that the octopus has had to develop so much brain power to keep it alive through such a rapid growth and development period. Janet Voight, from the Field Museum, explains that from what we know about an octopus's quick larvae-to-juvenile growth spurt, it must keep learning and updating its own information about what makes a good lunch, what is dangerous, and what might be a good place to hide. "As their size grows, what used to be food is no longer good enough, so they have to learn," she says. So to keep up with these ever-changing tasks it needs a good noggin.

On the other hand, Boal suggests it's not so much growth or a busy environment that has fueled the octopus's brain but, rather, a need to survive with a soft body in such a dogfish-eat-dogfish world beneath the waves. It had to be smart, she says, in order to cut it competing against vertebrates while being shell-less, spineless, and, essentially, jawless. When other mollusks could secure themselves in a hard shell, the octopus had to remember where its den was. Intelligence, in effect, has served as the octopus's defense mechanism. Biologist Andrew Packard supposes that without a protective shell, cephalopods gained an upper hand in terms of mobility, being able to make quick attacks or evasive moves—and squeeze easily into tight hiding places.

Roger Hanlon, who coauthored the book *Cephalopod Behavior* with John Messenger, was perplexed by the question: "Why *do* cephalopods have such big brains?" He and Messenger concluded that it is the wide range of behavior of some of "the best predators on earth" that they need to compete against—and avoid being eaten by. "You can't have this amazing diversity of behavior without something to organize it, control it," Hanlon says. And that means brains.

The evolution of our ancestors' intelligence from early tetrapods to modern primates seems relatively clear-cut—at least when compared to the puzzle of mollusks. With mussels and snails—which, as Boal points out, are "not exactly Einsteins"—in the same group, how did cephalopods get to be so darn smart?

To answer that question, some researchers are going back to the octopus's environment to see what lessons can be learned from comparing it with its direct competitors. "The fish in the modern era are sitting right next to the octopus, and they're dealing with the same world," says Frank Grasso, a biologist and engineer at CUNY in Brooklyn. "The fish had this essential vertebrate brain, and the cephalopods have this essential cephalopod brain, which goes back 505 million years."

These two types of creatures "were dealt a different hand from the genetic deck of cards," Grasso explains to me. And with the humble collection of starter genes that the cephalopods' ancestors received, "How do you build a large brain that's capable of complex behavior?" he asks. "What are the design principles? The vertebrate model isn't the only one, and if you look at the structure of the cephalopod brain—in particular the octopus—it is comparably complex, but completely alien in design. And yet they solve the same problems that fish solve and seals solve when they deal with the same environment."

A 1971 paper by Packard proposed that the millions of years of competition between cephalopods and fish had driven the evolution of cephalopod intelligence, making their behavior and possibly even their brain structures closer to fish and other more familiar vertebrates.

But what if it was the other way around? What if *cephalopod* smarts had been driving the fish to be smarter, more like *them*? That's what Grasso and his CUNY colleague and wife Jennifer Basil proposed in their 2009 paper "Evolution of Flexible Behavioral Repertoires in Cephalopod Mollusks."

He suggests that we should pay closer attention to the fossil record to see how the changing playing field in the sea might have contributed to behaviors we see there today. For at least 300 million years, "cephalopods were competing against extinct species of cephalopods—extinct species of big brain creatures," Grasso says. To survive, fish, too, would need to step up their cognitive abilities to contend with all of the highly evolved, big-brained cephalopods. So "the vertebrate brain, in fact, might have emerged as a reaction to the existence of cephalopod brains," Grasso says. Think about that next time you're chewing on some calamari. Perhaps we owe them some thanks for our own highly evolved intelligence.

To Think Like an Octopus

Despite all of the opened jars, dampened researchers, and decorated dens, we still have but a superficial sense of the depth of the octopus' psyche. And as landlubbers, we stand at a distinct disadvantage in wrapping our *own* brains around those of aquatic animals. An underwater existence requires a rather different framework for understanding the world than does a life lived in a forest or on a plain. We are still early in the process of learning what we don't know—and how to go about figuring it out. For example, how do they think about the world around them? *Do* they think about the world around them?

Jean Boal has heard this question too many times. "I almost think that I've been in this field too long," she says with a laugh when I ask her. "I no longer really know what the question means. I've come back full circle to thinking about 'Well, what does the animal *do*?' rather than trying to imagine what's going on in its head." For instance, she has been itching to see if octopuses engage in causal reasoning, the ability to discern cause and effect from unrelated occurrences. (Example: When you hear a car door slam, then your doorbell rings, and pizza arrives! But we, or most of us, know that if we slam a car door and ring a doorbell, that will not *cause* the pizza to appear—even if we very much want it to.)

But setting up that sort of experiment for an octopus is tricky. You have to teach it that one event—say, a flash of light—we'll call A precedes a piece of food coming from a hole in the right side of the tank. This food we'll call B. And that is followed by another event, say a burst of air bubbles, which will be C. So, logically, if it gets the blast of air bubbles (event C), it should check back and see if it missed the previous events and see if there is food in the hole in the right side of the tank. *But*, if there was a way the octopus could also trigger the air bubbles on its own, "it should say, 'well, *I* caused C; I probably didn't miss anything, and there probably isn't any food over there,'" Boal explains. "That would give us a clue—how an octopus could assemble information and use information." An experiment like this might work, but in order for it to be accepted scientifically, each step needs to be tested and validated to make sure that it's proving what scientists think it's proving. But if causal reasoning can be shown in cephalopods, it would give us a new way to figure out "what can they *do* with what they know?" Boal explains. And that, in turn, should give a hint as to what they *know*.

Jennifer Mather agrees that observing behavior might be the best method for glimpsing inside the octopus brain that we have for now. "We have to sneak around the periphery to show that what they *do* shows us how they think." We might be able to use navigational abilities to start to crack this quandary. "Intelligence means taking information from the environment," Mather says. "Their brains are not like ours, and they're paying attention to different factors, but intelligence is intelligence is intelligence." And now that we have more data about octopus spatial learning, we can start to look at how different species with different sensory systems accomplish similar things, such as navigation. "We have a wonderful opportunity to study how thinking works in many different animals," she says. Nevertheless, there will probably be things about the octopus that we'll never be able to know. "My guess is that it's going to be impossible to say any animal has consciousness," Mather says.

Of course trying to understand any other organism poses a grand phenomenological dilemma. As philosopher Thomas Nagel noted in his famous 1974 essay "What Is It Like to Be a Bat?" many animals doubtless have distinct experiences, and thus, in his estimation, there must be something that it is like to *be* that animal, such as a bat. But because the bats live in such a different sensory and cognitive world—using flight and echolocation in their day-to-day duties and possessing brains that are built quite differently from our own—"there is no reason to suppose that it is subjectively like anything we can experience or imagine," he writes. "This appears to create difficulties for the notion of what it is like to be a bat"—even, as he later notes, if "one spends the day hanging upside down by one's feet in an attic." After all, just by adopting an animal's outward behaviors, one's experiences will still "not be anything like the experiences of those animals."

Some researchers have been so won over by the octopus's impressive—if mysterious—cognitive abilities that they refuse to indulge in even a little octopus salad. Others have taken it to the next level and are looking to beef up considerations for them as research subjects. Although it might mire their work in more paperwork, some cephalopod-studying scientists have made the case for octopuses to be given special consideration as higher organisms. Although technically—and very obviously—they are invertebrates, in the UK, octopuses and other cephalopods have already been knighted as honorary vertebrates for research purposes, making them the only invertebrates legally protected from scientific experimentation that might cause "pain, suffering, distress or lasting harm." This distinction removes them from the experimental category of worms and fruit flies and into the regulatory tier occupied by rats, rabbits, and other nonprimate lab animals. The European Union has also launched a new set of rules that requires similarly vertebrate-like treatment of all cephalopods in scientific research. (In addition to adding procedural hoops, though, some people in the field point out that trying to lessen discomfort for octopus subjects could be questionable scientifically. We don't fully understand how the animals' nervous system and pain pathways work—and in what ways tampering with them might mess with the topics under study.) And the 2012 Cambridge Declaration on Consciousness mentioned them explicitly, along with birds and mammals, as possessing the requisite neurological framework for enough subjective experience to qualify as "consciousness."

To get a full grasp of octopus intelligence we need a broad and comprehensive search. An octopus's smarts aren't even entirely wrapped up in its head. In fact, aside from the optical lobes, which are stationed behind the eyes, most of its central brain surrounds its esophagus. And the majority of an octopus's nerve cells are spread throughout its body—approximately two thirds of which are serving duty in its arms. This diffused-neuron model is something even the U.S. military has taken a keen interest in. It has been funding research to see how individual tasks are performed and decisions are made at the local level, rather than having to bother the higher-up central brain with pesky tasks like manhandling a clamshell—or taking down an enemy.

CHAPTER SIX

Armed—and Roboticized

Octopus arms have been an inspiration for superheroes, supervillains, neurological studies, robots, and even for a particularly disconcerting Korean dish. These besuckered eight arms might be an octopus's most entrancing feature. They can bend in nearly every direction and extend to nearly twice their resting length. If lost in an accident, they can even grow back—suckers and all.

Rather than being controlled entirely by the octopus's central brain, each arm contains enough neurons to operate semi-independently. Even each separate sucker can extend, release, and pinch on its own. The arms might even be able to communicate among one another to coordinate tasks, tasks that can be extremely brutish, such as strangling sharks, or extremely delicate, such as removing surgical thread.

Roboticists see the octopus and its amazing arms as a playful challenge from Mother Nature. Not only is the animal enticingly soft bodied (unlike most robots today), but it also boasts this impressive intra-arm communication and diffused intelligence that could be useful for our own autonomous bots.

But it is the real octopus arms, with their multitudinous talents, that continue to beguile biologists and, as I found out, the adventurous eater.

An octopus is not one to go down without a fight. Even a disembodied arm, served up on a plate, doesn't give up the ghost quickly or easily. At Sik Gaek restaurant in Flushing, New York, diners can order a squirming Korean specialty: "live" octopus. I ventured out to the corner of Crocheron Avenue and 162nd Street after work on a balmy May evening (by way of the Number 1 train to Penn Station, the Long Island Rail Road to a wrong stop, to an incorrect address on Yelp, and at last, after a twenty-minute walk through a suburban neighborhood of Flushing). When you do make it to Sik Gaek to confront your outré culinary curiosity, don't plan on entering unnoticed. Each customer who arrives is greeted with cheers and hoots from the entire waitstaff—like a new wrestler entering the ring. But the uninitiated don't quite know the fight of the opponent they're about to face.

The restaurant's small wooden tables are low, surrounded by even tinier red and blue plastic stools. In the center of each table is a recessed burner to accommodate a hotpot. "With a little bit of oil and plenty of vegetables, let an octopus dance on the hot plate," the menu proclaims in big red letters. The fresh octopus hot pots will run you as much as $99.99, but it will easily feed more than four adventuresome adults.

I place an order instead for the $24 "fresh octopus 1 dish." I then head over to find the actual octopuses, which are sitting, rather resigned, in a tank by the open kitchen in the back. The manager, who goes by the nickname T, says that they easily go through twenty octopuses a day. For Koreans, this is hardly a novelty dish, he says. It's incredibly good for you—so much so that in Korea, farmers give it to their sick cows or bulls, and it will fix them up in a day, he explains. I'm no bovine, but I'm ready to give it a shot. The preparation for my dish is simple (if not for the squeamish):

Fresh Korean-style Octopus, or "Sannakji"

Courtesy of T at Sik Gaek

Take one live octopus, kill it instantly
Remove the organs from the body
Chop up the arms "sashimi style" into small sections
Serve on a leaf of lettuce with sliced raw garlic, green onions, and jalapeños
Optional dipping sauces: oil with salt and pepper or a traditional Korean hot sauce

Before I can ask T another question, I spot my dish arriving from the kitchen. Already?! But of course—no cooking involved (now that is some live raw food—if not exactly vegan).

On the plate, the muscular arm segments look like little slugs, writhing about, gray and stubby, seething all over one another and the garnishes. But my strange fascination and octopus mania has taken over any apprehensions I had on the train ride out here. I dive in.

The first one I pick up is little but squiggly—T crouches by my side watching me eat it. I dip it in the oil and salt and pepper sauce and drop it into my mouth, where it keeps wriggling around. The menu doesn't lie—it's the freshest octopus, and probably the freshest-tasting seafood I've ever had. It's slightly chewy, which is to be expected, but it's not at all as rubbery as some octopus sushi or grilled octopus I've encountered. It's, well, delicious. I go back for more as T gets called away to check on more tables.

This dish is also an exercise in advanced chopstick skills. Slippery noodles have nothing on the squirming, suctioning, muscular hydrostat of octopus arms. And despite their disembodied state, these suckers are still incredibly strong. Some of them I never quite can get unstuck from the plate. One sturdily affixed piece that I at last rip off the plate from its grip makes popping sounds all the way, like a tiny shower mat being ripped off the bottom of a tub.

But once you do manage to get one of these dang things on your chopsticks, it will likely wrap or suction onto the wood with one end, another end twisting around in the air, as if exploring like a blind inchworm. These innervated instincts don't stop inside your mouth. There, they continue to squirm—and, to my surprise, to suction. They will use their petite but powerful suckers to grab onto your gums until you overpower them with your own muscular hydrostat—your tongue.

T swings back by my table about five minutes later and seems surprised that I'm making steady headway on the dish. As some of the pieces start to slow down, they're a bit easier to examine. Their skin is a gray-white hue with fine, tiny purple specks. Perhaps a primmer diner might describe it as a death pallor, but I think it is beautiful.

Keeping rough track of the time, I see some feisty pieces still going strong at least fifteen minutes after my dish arrives. One saucy segment had latched firmly to the plate as well as to a slice of raw garlic. I finally get it off the plate, but the garlic comes along too. Even some of the slower-moving pieces seem to be reanimated by a dip in the sauces. The salt provides sodium ions, which trigger still-firing neurons to spur muscles into action.

Research has backed up what Korean cooks have known for ages: An octopus's arms can move on their own accord for many minutes after the physical nerve connections to the main brain are severed. This is possible because each octopus arm contains its own array of nerve cells dedicated to operating somewhat independently of the main brain. So although this ability makes for a somewhat disarming dish, in the wild, it helps the octopus scour crevasse-filled surfaces for its own dinner.

Not everyone is such a fan of this culinary tradition, however. PETA is not exactly pleased with the idea of serving "live" octopus. It has called for letters of protest to be sent to a local district attorney and even staged a demonstration demanding the restaurant stop serving the dish.* Many American diners seem generally creeped out by their encounters with this dish. But strangely, as revolting as it might look to the amateur eater, the dish left me thinking about it and, dare I say, missing it for days afterward. It was the most intimate eating experience I've ever had. Although for the poor octopus it was not the best of times, to me, it felt almost as if we shared the dining experience.

Minds of Their Own

Even when we can't see our arms and legs and aren't giving them much thought—say, when kicked back in a hammock, gazing absentmindedly up into the blue—we have a pretty good sense of where they are, whether they're straight or bent, and more or less what they're doing (swatting at flies or kicking off sandals).

* PETA has also made its displeasure known about the Detroit Red Wings fans' decades-long practice of throwing octopuses (live, stuporous, and dead) onto the ice at games during the Stanley Cup playoffs. The significance for fans is that the playoffs once required eight wins.

For all of an octopus's apparent smarts, its central body awareness has been presumed to be worse than that of a gangly teen in the midst of a growth spurt. An octopus is considerably more coordinated—and certainly more graceful—than a thirteen-year-old boy on a soccer pitch. When this clumsy kid comes clamoring down the field, it might appear that each of his newly long legs has a mind—a not so bright one—of its own. Whereas such an unkind characterization might be working against the awkward adolescent, the cliché holds quite true in an octopus—and actually helps it out. Each of an octopus's arms is capable of carrying out general instructions from the central brain and then, seemingly, of improvising as well. Scientists suggest that this is an evolutionary imperative, as with practically an infinite range of movement along each jointless arm, the amount of brainpower needed to stay on top of all of that unwieldy information would be rather unreasonable.

Although just about infinitely flexible, the octopus arm can also be transformed into a vertebrate-like appendage, complete with specific, finite segments like our own arm. A 2006 study showed that these bendable animals sometimes fixed their soft arms into a more stiff-jointed extremity, creating straight segments between three "joints." This elbowed arm might help the octopus keep track of its motion by giving it a finite range of movement, the researchers suggested. It also suggests that a three-jointed arm, like the ones we are blessed with, is a pretty good evolutionary solution.

A clam-cracking octopus is working with something much more complex than the phenomenon we four-limbed humans know as muscle memory. When we perform a familiar physical task, such as riding a bike or dancing the two-step, the process can feel so automatic that it might seem as if our arms and legs are working of their own accord. But we're really sending out specific—albeit well-practiced—instructions from the brain to different parts of the body. The brain, in turn, is getting near-constant feedback from our arms, legs, and other inputs (Are the handlebars stuck? Did I just two-step on my partner's toes?) and is adjusting its outgoing messages accordingly. But what if we could skip that process and outsource much of the intelligence analysis to individual body parts?

Each octopus arm has a chain of nerve bundles, or ganglia, that run up the center of each arm. Binyamin Hochner and his team at the Hebrew University have discovered that simple, frequently used arm-reaching actions are controlled by the nervous system of the arm itself. These, in turn, are hooked up to a circular bundle of nerves just above where the arms split. So in addition to some sweet independent action, it seems that the arms can get in touch with one another without having to go through the central brain. In this way, they can cue adjacent arms to help out with a task if needed.

The formal scientific description of this odd autonomous anatomy might be relatively recent, but the notion has pervaded cultural understanding for more than a century. The octopus has long been the symbol of disparate but powerful organizations, such as the Southern Pacific Railroad monopoly in Frank Norris's 1901 novel, *The Octopus: A Story of California*, and the Italian mob in Claire Sterling's 1990 nonfiction exposé *Octopus: The Long Reach of the International Sicilian Mafia*. The animal also lends its name to a conspiracy theory propagated by Cheri Seymour's 2010 *The Last Circle*, which is taken down in Guy Lawson's 2012 Wall Street investigative thriller *Octopus*. And on the populist side, it has been invoked to describe writhing mobs of protesters and the satirical "Octopi Wall Street" meme.

Even consumer tech products frequently allude to the animal's long, useful arms. The Hong Kong mass transit system rolled out the first contactless e-card called the Octopus Card in 1997. Known as the Baat Daaht Tung in Cantonese, it translates loosely as the "reaching everywhere pass." Cardholders can now also use it to make purchases at participating stores, pay at parking meters, and even check in at school.

Octopus-inspired characters, such as Spider-Man's nemesis Doctor Octopus (aka Dr. Otto Gunther Octavius or Doc Ock), played by Alfred Molina in *Spider-Man 2*, often make use of seemingly autonomous arms. After a freak accident (shocker!), the highly intelligent, once shy and reclusive Doc Ock gets mechanical arms fused to his body—and brain. Even when they are eventually removed, they retain the ability to move on their own, performing handy tasks one might want from such independent extremities, such as coming to break him out of jail. These well-articulated arms could also split up tasks among themselves, some employed to fight off opponents while others engage in a more delicate task, such as fetching something to drink.

While authors and marketers are happy to run with the metaphor, biologists, engineers, roboticists, and the military are still trying to figure out how all of the somewhat independent arms work together and apart. In actual laboratory work, these efficient communication tactics inspired the U.S. Department of Defense to fund research on this distributed intelligence system. The parallel to hierarchical command structure, like a commander over troops, is a tempting target for military research.

All of this is made possible, scientists think, by the octopus's diffuse nervous system, which can be a bit of a stretch to relate to. But there is one familiar example that biologist James Wood suggests can help us better understand octopus actions: the reflex test. Think of that unpleasant little pointed rubber hammer doctors tap on your kneecap that ideally sends your leg jutting out, embarrassingly uncontrollably. Although that's just an automatic reaction, it's one small realization that even *our* bodies are not entirely controlled by our over-thinking cortexes.

"You have this big CPU in your head, and it does almost all of the processing," Wood explains. Octopuses, on the other hand, are "more like the Internet," Wood points out. The Internet is not one central command center, but rather it is made up of multiple nodes, gaining stability in its redundancies. If some of those pieces go down, the Internet as a whole can still function. As cephalopods have reduced the need to rely on a central brain to execute their many movements and decisions, so have we been spreading our data and computational load off our own individualized CPUs and distributing them to decentralized processing in the cloud, making for a more efficient—and reliable—network. So as we humans move away from machine-based computing and into the cloud, we're making one small step toward creating a more cephalopod-like system. "It's all from the same problem, but with different solutions," Wood says.

Regrowing Limbs

Not only are octopus arms relatively autonomous, they are also easily replaceable. Although they're not harvested individually, these impressive arms are technically a renewable resource. Like lizards that can grow a new tail or starfish that can grow a new limb, octopuses can regrow innervated arms if they are lost. This regenerative ability was first described in scientific detail in 1920 by zoologist Mathilde Margarethe Lange. A U.S. biologist who studied in Germany, Lange built upon field reports of partially regenerated appendages and detailed in a paper in the *Journal of Experimental Biology* the process of octopuses' regrowing arms while they were in captivity.

When an octopus loses an arm, whether to a predator or an accident, it can close off the affected arteries to avoid excess blood loss until a small stub starts to form and sprout a new arm. This process is apparently efficient enough that some octopuses will voluntarily cast off an arm if threatened—like a lizard its tail. And thanks to the substantial independence of its nervous system, the severed arm can move around on its own and change color, possibly serving as a decoy for a predator, while the seven-armed octopus attempts to make an escape.

The regenerative reflex is also under study by scientists trying to figure out how we humans might be able to grow back, if not whole limbs, at least important nerves lost to disease or injury. It might seem far-fetched to try to emulate such anciently evolved invertebrates, but Heather Bennett, of Illinois College, has pointed out that some of the same genes (known as the *Hox* genes) that regulate initial arm formation in humans are those that do the same in octopuses, a link that could help inform regenerative medicine research. If scientists can figure out how the octopus versions of these genes differ from those of humans, we might be able to make similar tweaks to human *Hox* genes in the lab. These could encourage the regrowth of nerves to fix those in people with neural injuries or diseases.

Amazingly, when an octopus regenerates an arm, it grows back with suckers and everything. And these suckers, it turns out, are far more complicated than they look.

Smart Suckers

After the eight arms, one of an octopus's most striking characteristics is its multitudinous suckers, each of which is perched atop a flexible stalk and can extend, bend, and even taste on its own. A common octopus has about 240 suckers—per arm—which adds up to a total of 1,920 to coordinate.

Octopuses are not alone in having developed such an attractive solution for grasping. Some bats and fish also use suction-based gripping techniques. Cephalopod suckers were, however, the inspiration for suction cups—both the ancient variety, made from gourds, and the modern iteration, patented in 1882.

But an octopus's suction cups resemble the familiar flexible imple-

An *O. bimaculoides* displays its suckers.
(Denise Whatley/TONMO)

ments we use to stick tacky thermometers to our kitchen windows only as much as Navy SEALs resemble plastic toy parachute men: they are roughly the same shape.

Not only do octopuses not need spit and a prayer to make their suction cups work, they can engage and disengage them at will: no pulling, peeling, or cursing required. And these suckers far outdo those of their cousins the cuttlefish and squid, whose versions are mostly just meant to grip—if that. The octopus can move and rotate each sucker individually, and even fold them together to use as pinchers. "It turns out that it's horribly complicated," Jennifer Mather says. To orchestrate this subtle ballet, the octopus possesses an entire, separate neural ganglion to control just the edges of their suckers.

In their book, Mather and her colleagues described watching one octopus exhibit its extreme sucker skills seemingly without a second thought: "Once when we watched a Hawaiian day octopus resting in its den, a tiny foraging hermit crab fell on it. Without moving its coiled arm, the octopus picked the crab up with a single sucker, extended the sucker stalk, and dropped the offender a bit farther away."

And when an octopus wants to eat something it has picked up with an arm tip but doesn't feel like moving its whole arm, it can pass a small object along the length of the arm using only its suckers—a move Mather and her coauthors call "the conveyor belt."

These fine movements might actually be, rather than a recent evolutionary development, a feature of early cephalopods as well. A 300-million-year-old cephalopod fossil preserved enough detail around the suckers to convince Frank Grasso that they had also had these extrinsic muscles.

In addition to manipulating objects, octopus suckers are constantly gathering information about their environment, including tactile and chemical signals. This helps the animal to figure out whether something it has in hand, so to speak, might make a good lunch. Each sensor has neurons that collect and distribute information from the suckers' chemoreceptors (tasters), mechanoreceptors (pressure sensors), and proprioceptors (position detectors).

In his lab in Brooklyn, Grasso has been studying the subtleties of the octopus sucker in hopes of co-opting some of their natural abilities to use in artificial intelligence–equipped robots. He and his team at the Biomimetic and Cognitive Robotics Lab at Brooklyn College are probing the ways in which, like the arms, octopus suckers seem to be able to coordinate movement and activity with their neighbors independently of the main brain.

But to study an octopus's suckers, first you have to get it to hold still—which is no small task. So it seems perfectly fair for Grasso to brag a little that they have developed a method for safely anesthetizing an octopus and keeping it still in a homemade restraining device. "It becomes fully awake when we're doing these tests—and doesn't struggle against the restraints," Grasso says. "It just kind of hangs out there and doesn't seem to show any ill effects." On a flat sheet with its suckers upward, the octopus is ready for study.

Each sucker's ganglion has a lot of brainpower. So much so that Grasso calls each of these nerve bundles a minibrain. That sucker minibrain has thousands of neurons, which link up together to the larger line of nerves at the center of each arm. For each sucker, "that's a lot of processing power," he says. They have been looking for ways to record what is happening at different parts of the nervous system when the suckers are stimulated.

Studying the path of information from the suckers, Grasso notes, also helps to "map the nervous system." He compares the whole operation to that of a smart power grid, or a well-connected network. "When you think about it as one node in a network—like a social network of people—then they're all communicating with one another up and down to coordinate their activity," he says.

Grasso and his colleagues have been studying octopuses' reactions to different types of chemosensory information fed directly to their suckers. To give octopuses a "taste" of something on individual suckers without giving the octopus a distinct physical sensation is difficult. A blast of something tasty would be too obvious, but a light waft of it over the water would be too diffuse. So Grasso's lab has been using flavored ager gels. Some were mixed to be an acidic or a basic version of seawater; another was ground-up fish—pureed down to the amino acids with the solids removed, "just like making a perfume," Grasso explains—and a final flavor: ground-up octopus.

By giving octopuses these gels on the individual suckers, they were able to track just how different the response was to the various flavors. "There are tremendous differences in the reactions to octopus compared with acid and base," Grasso notes. They also found that the sensory input (chemical and mechanical) seems to stay separate far into the internal wiring system. "They don't get mixed together in that individual sucker processing center, like the sucker ganglion or the brachial ganglion," Grasso says. "They appear to be at least preserved as separate streets," he says, which provides clues about how octopuses are processing—and experiencing—these stimuli.

The suckers are not only sensitive and smart, they are also strong. They actively attach one by one. And even postmortem, with the help of a little liquid (olive oil or vinegar will do) dead octopus suckers can still form a seal, like those in an octopus salad I had to fight for bites of in Greece. One tiny quarter-inch sucker can hold more than five ounces. And some researchers have figured that just a single large sucker can hold thirty-five pounds or so. So that's one powerful grasp.

This makes them tough to pull off skin too, as actor Mark Wahlberg discovered when he brought his young son Mickey to meet an octopus at the Sea Life Aquarium in California. "The thing latches on to my son's arm," he recalled later on *Late Night with David Letterman*. "I was like, 'This is calamari. I will kill this thing,'" he said. "Finally, we rip the thing off of him. He's got all these suction marks that look like about thirty little hickeys." Plenty of researchers are also quick to warn about these embarrassing marks.

A sucker's strength depends in large part on how much volume it holds. Under water, that force is limited by the weakness of the water molecule itself, Grasso explains. Once a sucker is stuck onto something, "if you reach the point where water capitates—where you're actually breaking apart the molecular structure of the water—so it is not holding itself together anymore, it will break," he says. So in theory, "the octopus could generate more force of attachment if the water" itself were stronger. Grasso and his team have been able to re-create this in their own work above water. "The same principle holds in air," he says. "The robot suckers we built were able to hold masses up to the theoretical limit set by the volume of air inside it."

Grasso and his colleagues are also studying the exquisite control of these suckers to try to replicate some of the capabilities in robotics.

Our New Octo-overlords

You knew it was coming. And yes, they're almost here. The octopus seemed destined to be the next autonomous robot. (Let's hope they end up working for good and not evil, like a Robocopoctopus.) To move the robotics field from the era of the hard-shelled, fix-jointed *Star Wars* C3PO-like model to something a little bit more, well, flexible, teams of engineers and programmers have taken a liking to cephalopods.

Grasso qualifies this work as "the concept engineering" inspired by solutions the animals themselves have developed. His team has been constructing a robot octopus sucker that contains sensors akin to those on a real octopus. It would know if it was in physical contact with an object, "and then it would form a seal," Grasso says, "and then it would attach and hold on to it." Thus it would begin to mimic a real octopus sucker. Perhaps someday they will also be able to include more subtle sensors to detect chemical signals too—even if it isn't (one hopes) looking for lunch around the lab.

One of the largest challenges that engineers face in developing robotic cephalopods are the materials. "To really get what the biological systems can do, we need some real innovation in materials science and actuation," Grasso says.

But none of these obstacles are insurmountable. Why? The solutions already exist! "The octopus, the squid, and the cuttlefish prove it's possible. These things physically exist," Grasso says. "Animals are existent proofs for problems that we need to solve. And the question is just connecting the dots to do those things."

Materials aside, there are also some big questions about integrating the sensory with the physical. "Simply understanding what it is that cephalopod nervous systems and bodies do as they interact" is one of the keys, Grasso says. And that might take a little reverse engineering.

Grasso agrees with many of the other engineers in the field that there is no need to tarry until the biological research has caught up. "We shouldn't wait at all. We should work on everything all the time!" he says excitedly. And engineering can also inform the bio side. It's possible, he notes, "that engineers working on developing loosely inspired cephalopod models would be able to make some kind of insight or an advance that would simplify our understanding and inform our biological studies. So it can go both ways."

In fact, he sees his own work that way: "Building octopus-inspired robots is a way of understanding what their brains are doing," he says. And that holds true for all of the animal-inspired robotics projects he has worked on (which also include robot lobsters). "If I wanted to build robots, I would just be building robots, and I wouldn't be yoked to the biology. For me the robots have always been a way of testing hypotheses about the animals—particularly their brains. My goal, really my overarching goal, is to understand this brain architecture that is so alien," he says. "These animals are so sophisticated in their behavior—in their abilities. And robotics is just a tool to help me take a few pieces from that," he notes. "But I have to step back from that, because I know I can't possibly get there," he says. "There are not enough years left in my life to do that."

Another team, meanwhile, is working on its own smart and strong robot octopus arms. And no, this is not the Spider-Man comic book team. This is real research, backed by serious grant money from the European Union.

The self-proclaimed OCTOPUS Project* has been hard at work on this task. This multi-institution, international collaboration is working toward a fully autonomous robotic octopus that could, like a real

* Not to be confused with the four-person—and eight-armed—Austin, Texas–based indie band Octopus Project.

octopus, accomplish feats that no hard-jointed bot ever could. Cecilia Laschi, a biorobotics professor at the Scuola Superiore Sant'Anna in Italy, has been coordinating the effort. They demonstrated a prototype disembodied octopus arm, and they're now working toward the remainder of the body, from mantle to suckers. Their goal is to create a robot that will move like the octopus underwater and, like the real deal, be able to reach into tight spaces. It could be used for search and rescue as well as for exploration. It's also in the works simply to prove that it is possible to create an entirely soft-bodied robot.

I went to visit Laschi and her colleagues at their lab in Livorno, a short train ride outside of Pisa, on the Tyrrhenian coast. Their small, modern facility is on a locked pier next to a grand, crumbling bathing house built in 1846 for the queen of Eretria. One of Laschi's researchers, Laura Margheri, meets me at the imposing nineteenth-century trident-topped gate. This Poseidonian barrier hasn't seemed to deter the ardent Italian sunbathers, who have climbed around it to lie near the water. Do they know what crazy futuristic tinkering is transpiring in the unassuming building nearby?

A bubbly biorobotics PhD student from Florence, Margheri leads me into the small, open lab, her tight curly hair bouncing as she walks and talks. Inside, the many side doors are flung open, letting in a gentle sea breeze and a view of boats in the harbor (a vastly different setup from most labs I've seen, which are often relegated to university basements or sealed-up upper floors). This lab has the happy, semichaotic feel of in-the-thick-of-it science. At rows of workstations, some graduate students and postdocs are working away at their computers and tinkering with prototypes. Off to one side is a half-full inflatable kiddie pool. And in the center, at the head of the lab, is a large, well-appointed saltwater tank complete with rocks, starfish, and one aging—but active—octopus. The lab members tell me his unofficial name is Andreino, after former lab mate and colleague Andrea, who had caught it. When a few researchers gather around for a photo in front of the aquarium, Margheri tries tapping on the tank to get the octopus's attention so it can be in the picture. When that fails, she reaches in with a long grabber, which

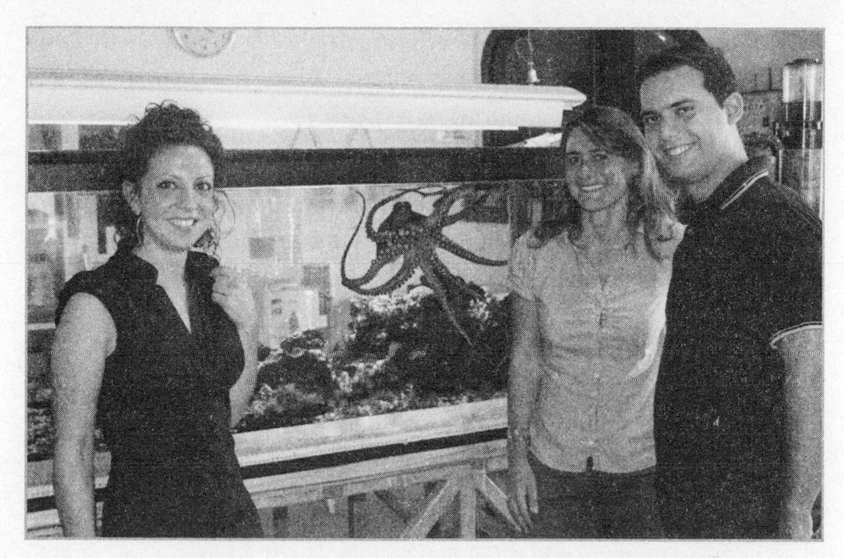

Margheri (left), Andreino the octopus, Laschi,
and Matteo Cianchetti pose for a picture.
(Katherine Harmon Courage)

I assume will make it dart away into hiding. Instead, she succeeds in alerting it so it, now sees everyone assembled on the other side of the tank and swims over, latching longingly onto the glass for the photo. This old octopus doesn't seem antisocial at all; it just wants to be part of the gang.

Margheri has been working on testing the octopus's natural abilities in hopes of mimicking them in the robots. For example, she has rigged some clever experiments to see just how far octopuses can stretch their tonguelike arms. In one exercise, a common octopus learns to retrieve a treat from the end of a long tube with one of its arms. Learning this usually takes just five sessions over a period of a couple of days. She can then extend the food farther down the tube and measure just how far the muscular hydrostatic arms can stretch from their resting length. The arms, it turns out, can extend to about double their original length—an engineering challenge indeed.

Although Margheri comes from a bioengineering background, she says she enjoys working with dedicated biologists, because they get excited using some of the precise engineering tools to measure the animals' capabilities. However, as she points out, the live-animal octopus experiments have been a bit difficult. Most types of lab animals will tolerate (or not notice) small dots stuck on their bodies or limbs for researchers to track movements with high-speed cameras. Octopuses, however, will remove everything in short order. "It's not easy, I can tell you," she says with a laugh. Like the behavioral researchers, Margheri and her colleagues are used to finding all kinds of broken instruments and tubes in the octopus tanks.

The project leader's own background is in more traditional robotics. "I'm used to robots that have rigid links," Laschi explains. After working with neuroscientists and learning more about our own brains and how we coordinate our bodies, she says, she started to feel a little frustrated by the traditional rigidity of classic robots and the absence of structures like muscles. So she and some more bio-oriented collaborators started to plan a daring soft-bodied robot project. And what better model than the octopus? "All biological systems have some soft material," she says, "but the octopus is very special because it has *only* soft material" (except for the beak, of course). "So we took it as the extreme—if you study this end of the spectrum, you can solve the others."

The team has been able to get a rare inside view of the stretching octopus's arm. With ultrasound technology they have been able to see the muscle and nerve structures at work. This rare view helped to clarify "the secret of the movement of the octopus arm," chimes in Matteo Cianchetti, another biroboticist in Laschi's lab. In the absence of a skeleton, three groups of muscles give the octopus arm its flexibility and its structure, allowing it to change direction, length, and even stiffness. To try to replicate the muscles, the researchers are using cables and springs forged from shape memory alloys that can be bent and then returned to their original shape after being heated by an electrical current.

The muscles themselves are extraextendable, but the central nerve cord in each arm is not. Instead, it's arranged in a zigzag configuration down the arm, Margheri points out. This shape gives it extra room to straighten and stretch out, allowing the arm to extend without tugging on the nerve bundle. The researchers are taking a cue from the actual octopuses and are packing processing wires into the center of the arm in a wavelike pattern so they, too, can extend out like an octopus's nerve cord, when the arm stretches.

Cianchetti demonstrates one of the silicone-skinned prototypes, which has a ghostly gray hue to it. By pulling on a few wires, he can make it curl up organically, like a thick, slow-motion lasso. I stick my finger out, and it firmly wraps its strong, cool, rubbery surface around with a disconcerting ease. (Add this to the list of octopuses I would not like to meet underwater—or anywhere, really.) Just by virtue of the shape and proportions of the arm and the "musculature" inside, it naturally spirals around whatever it's grasping. "It automatically adapts to that shape," Cianchetti says. *Great*, I think with a shiver.

Along the robot arm are circular suckers, which don't yet work like actual octopus suckers do. The scientists are using these points along the arm as sensors to integrate tactile perception. Sucker manipulation will come later.

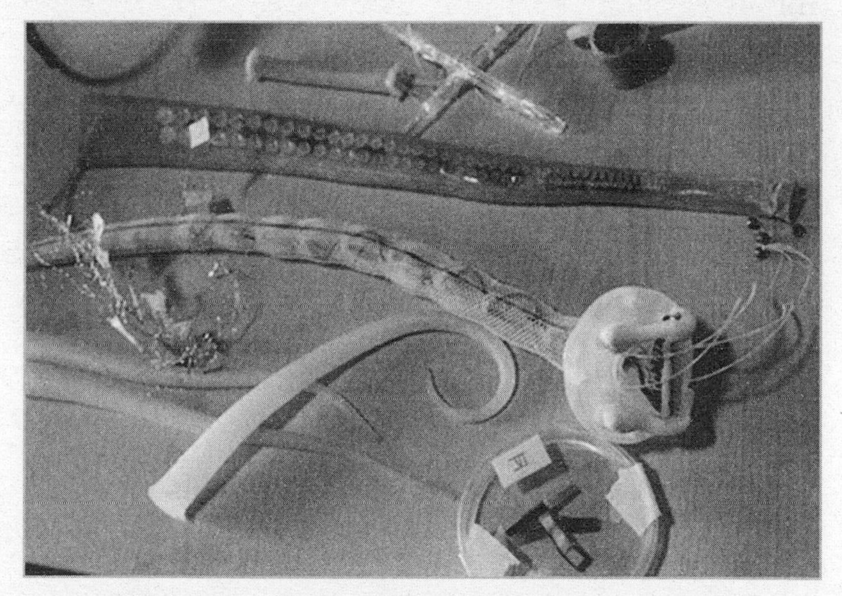

Components for a prototype soft-bodied robotic octopus arm.
(Katherine Harmon Courage)

These prototypes are steps along the way for honing the capabilities of the arms as well as the materials. Ultimately, the materials will have to be carefully chosen so the robot can perform well underwater for long periods of time and not rust or corrode. The silicone that they're using is very close to the density of water, so it is buoyant—just like a real octopus.

But a really cool underwater robot is much more functional if it can cover some ground, so some of the lab members are working to make it mobile. Marcello Calisti, another bioroboticist, is working on the walking problem. Most real octopuses seem to walk more with their back arms and feel around with their front ones. But for the artificial version, the roboticists might instead have it reach out with its front arms, attach some suckers, and pull, which will also help with exploration and determining directional movement.

Calisti has his workstation by a glass door that opens out to the harbor. He's also right by the inflatable pool that they fill with water to test underwater crawling and other tasks. Calisti shows me his current prototype, which is made out of hard-material motors and fixed cables, has only six arms, and still looks more like a robot a high schooler might build for a science fair. But there's something eerie about it when Calisti shows me a video of its crawling along, spiderlike. "It's a little bit creepy," he admits. So far they have been able to program it externally, turn it loose in the pool, and watch it locate and retrieve objects. But the goal is to eventually put the command center inside it (and give it the full eight legs and a totally soft body).

Octopuses, of course, don't use their arms for all of their travel. In the real world, octopuses opt for jet-powered swimming when they need a quick escape from a predator and to cover some serious ground, so they fold their arms together and take off. And you might not want to wait on a search-and-rescue robot to pick its way, arm reach by arm reach, to a particular underwater site.

At the other end of the small lab, Francesco Giorgio-Serchi is working on a side project to try to re-create the octopus's propulsion system in silicone. The water jet that octopuses can propel themselves with creates a swirl of water known in the scientific field as a vortex ring. The real animal generates this force by sucking water in and, using its mantle muscles, squeezing it out through the funnel. Scientists are just starting to learn more about this feat of physics, which is also employed by squid and select other underwater animals. It might seem simple, but we're only now figuring out the fluid dynamics it employs. The goal is to adopt it to propel small subs or autonomous vehicles someday, Giorgio-Serchi notes. As he points out, adapting this aquatic locomotion to our use would be a big step forward. In our primitive technology, most "every kind of propulsion in the water environment is continuous," he notes. Propellers and even water jet boats generate a constant motion. This octopus-inspired propulsion "would be the first example where you actually use a discontinuous jet." And not just for novelty's sake (although, let's be honest, it *is* pretty cool). "It's interesting," he says, "because it appears that it is especially efficient." Recreating a vortex ring could give underwater vehicles extremely strong and efficient acceleration.

But you can't just tie on a big turkey baster, fill it with water, and squeeze. The octopus's system is a bit more finely tuned than that. "Certainly the most complex aspect here is reproducing this capability of his to just contract a little bit, the width of the mantle, and change significantly the volume inside, which gets displaced," Giorgio-Serchi says. "It's a challenge."

The octopus, however, is doing this with ease. So Giorgio-Serchi, who has a PhD in geophysics, decided not to reinvent the proverbial wheel. Instead, he took a cast of a real octopus mantle and then reconstructed it in silicone. He shows me the detailed model. There are even cavities where the organs go, which, for now, he has filled with electronic components. "It's a big approximation," he concedes. But the results should also help inform biologists about how these cephalopods are swimming. Giorgio-Serchi notes that they're still trying to figure out all of the steps during an expulsion and how the octopuses are using their funnel, "which is basically the nozzle," he says, to control the output of water. This control certainly helps them to navigate, but it might also play a role in determining the velocity and flow of the mysterious vortex ring itself. During this type of swimming, an octopus will often flatten its arms together, streamlining its shape. Because it swims mantle first, its arms come together in a smooth bundle to get good distance. But they also might use their arms and web to help steer as they glide through the water column, so researchers will eventually need to balance these robotic limb components with that of the propulsion dynamics.

The next step for the robotic octopus will be to integrate some flexible intelligence as well. Aside from the engineering challenges, Laschi and the rest of the OCTOPUS Project researchers are vexed by the biological question "How can such a simple animal control such a huge amount of physical freedom and sensory data?" She is convinced that in the octopuses' case, "they cannot control everything with such a small brain." Obviously the jury's still out on how the animals do it, but that's not going to stop the engineers. So Laschi has two words: embodied intelligence, which means that each part of the body—octopus or robot—is, at least in part, in control of itself.

To run all of those arms so exquisitely, "there *must* be a lot of embodied intelligence," she notes. "Each arm has many neurons and controls a good part of the movements, but we don't have a real model from neuroscience of how it works really," Laschi says.

Not only does neuroscience fail to provide useful models, but traditional robotics also comes up short. Robotic control has been based on rigid, finite movements. But what do you do when you have a near infinite range of motion with multiple parts? This is, of course, precisely the question that biologists have been coming up against as well when looking at the octopus itself.

So as Laschi and her team press onward, they turn to a very basic biological principle: learning. Just as we—and surely cephalopods—learn at a young age how to control our limbs, so too will these soft-bodied robots. This will create the embodied intelligence she mentioned—and it will also do away with the need for exhaustive modeling. With this application, you could give the robot octopus a movement and then it will learn over time to adapt it for other tasks. For instance, if it encounters an obstacle, it might run through a variety and combination of various locomotion and arm movement commands. Once it finds the movement or combination of movements that work to surmount the impediment, it should be able to retain that lesson and then deploy the technique when faced with a similar obstacle. In this sense, it should learn somewhat as we do and become more "intelligent" over time—even if it might not ever be able to make the cognitive leaps required to build humanoid bots.

To get there, though, the team first needs to figure out body feedback systems, such as sensors in the arms to detect how much they are extended or contracted. They might be able to use the shape-memory alloy spring itself as a sensor. "We will have both tactile sensors and some kind of position sensors," Laschi says.

But whatever creepy, crawling robot comes out of the project, the researchers have not yet decided on a name for it. Conventionally, they would name the prototype for the name of the project, but that would just produce something like Octopus Octopus (or, to get scientific, *Octopus octopus*). So that part of the project, at least, might have to go back to the drawing board.

All of these robotic explorations are helping scientists better understand how octopuses themselves might control their impressive appendages. But, as with skin color control, we still have a long way to go on both the engineering and the biological sides.

The full capabilities of octopus arms will likely continue to flummox us for years to come. The octopuses, however, seem unconcerned, and are busy using their amazing arms to grab their next dinner.

CHAPTER SEVEN

Hunting

The light is filtering in from above. It's quiet except for the sound of my own breathing. Something bright catches my eye. I look left and spot two shockingly yellow fish. But as I'm admiring them, I suddenly realize—too late—that I'm being pulled by an unseen force to the right toward a big rock. I try to reverse course, but I bash my knee on an outcropping of coral that I hadn't seen below, causing a cut that aches for days.

Even down here, on a calm spring morning in the shallows of the Caribbean, the ocean is a chaotic place—a place utterly unlike the one we're used to ashore.

Just a couple of brief snorkeling expeditions in Puerto Rico to look for octopuses were all I needed to realize just how poorly equipped we are to keep up with these animals. Getting familiar with a subject and its hunting practices requires a challenging attempt to understand how they sense their world. For example, with a little effort—and a lot of patience—we might be able to spot an octopus out on the prowl in its natural environment. But that same landscape is going to look and feel and sound very different to the hunting octopus.

In this strange world, however, the octopus is a cunning and able predator. Unlike many other bottom dwellers, the octopus makes sure most of its meals are caught live. Using advanced vision and exquisite touch and tasting abilities, the octopus is a formidable foe for would-be prey. With strong arms and a toxic bite, it can feast on some of the toughest-to-shuck bivalves. So focused on food are they that some will even venture out onto land to find a meal.

But to learn more about how the real animals capture prey, one must visit their world to look for *them*—if only to observe. And so I soon found myself on a coral reef in Puerto Rico, on the hunt for octopuses.

Hunting for the Hunters

Roy Armstrong spotted it first. A pristinely clean clam shell lying near an outcropping of coral. "An octopus is nearby," he proclaims as we both surface and take out our snorkels. We go back down for a closer look. Armstrong expertly dives down and peers into all the perfect hiding spots, and I circle around, checking the surfaces of the coral, rocks, and substrate. I try to recall all of the octopus camouflage videos I had seen, but swimming down here in the Caribbean, with all of the brightly colored fan coral, tropical fish, and menacing urchins, it actually seems like a three-pound octopus should be easy enough to spot.

Or not. I should have remembered Roger Hanlon's words from the previous autumn in Woods Hole: "You can't go out on a reef and expect to find an octopus," he had said. "That's nonsense—I've done it wrong so many times, I just know it's nonsense."

I, however, had taken no heed and travel down to the southern coast of Puerto Rico in pursuit of these clever hunters. I arrive in the sleepy fishing town of La Parguera, where Armstrong has helped Hanlon and others find local octopods. But first, to find Armstrong you must drive all the way through little La Parguera until you come to the end of the road. Turn, and keep going until *that* road ends. You'll find yourself in a dirt parking lot. At the end of the lot, a *botero,* a man in a small boat, will be waiting to shuttle you across a narrow chute of water to Magueyes Island. If you pick your way over the small island's massive, lounging iguanas (the many descendants that survived from the island's previous iteration as a zoo), you'll find Armstrong in his impeccable office, perched on a yoga ball, wearing Birkenstocks, and pouring over satellite images.

Armstrong takes his university boat out for a water-sampling cruise several times a month. And having been in La Parguera "forever," as he puts it, he is intimately familiar with the reefs and their nooks and crannies (so much so that he corrects his boat's course for errors and omissions he knows are in the GPS maps). His shaggy brown hair, slightly crooked teeth, well-worn short-sleeved cotton shirt, and calm demeanor might lead you to think he's a laid-back beach-bum type. He's not. He's an exacting scientist who will spend upward of ten minutes finding the best spot to anchor his boat.

We meet early one day as the chickens of La Parguera are still cooing into the clear March morning air. He gathers our snorkeling gear and patiently fixes the sputtering outboard motors. And we are off to hunt for octopuses. It's a world away from the rough seas faced by the Spanish fishermen—at least today.

The breeze picks up when we reach our first snorkeling stop, near a reef called Media Luna. There are tons of fish swimming everywhere (big yellow snappers, schoolmasters, and blue tangs) and black urchins lurking in crevasses. But I don't see a lot of promising octopus chow.

We pick up anchor and move on to the next spot, Cayo Enrique. After a few minutes, Armstrong swims over to get my attention to show me the freshly cleaned clam shell. "There was an octopus around here

A clear day for snorkeling off the coast of La Parguera.
(Katherine Harmon Courage)

somewhere for sure," he says. But even with our thorough search, no cigar. (Although it's likely that there was an octopus there—and that it was keenly aware of *our* presence.)

Our third attempt, near a mangrove forest, didn't even yield any evidence. And another windy afternoon search off the northern coast by San Juan with Josh Rosenthal from the University of Puerto Rico the next day turns up even fewer clues.

Stumped though we were, our search tactics were not misguided. As Mike Henley, a keeper at the National Zoo, told me, when he goes out diving for coral research, he keeps an eye out for "Ringo's garden" as he calls it—any suspicious concentration of discards. The octopus's garden is often actually a graveyard of shells. These collections of carnage can help divers spot octopuses. They can also help researchers figure out what octopuses eat.

Judging from the piles of discarded shells that can accumulate in front of a den, the octopus is not a very picky eater. Some individual octopuses have been known to eat dozens of different species of prey, and others might just be limited by what's available in their area. Some species do, however, seem to turn their noses up at a meal that another octopus would snatch up in an instant. And some members of the same species, living in the same area will occasionally select different diets. (Humans are not usually on the menu, although Victor Hugo's novel *Toilers of the Sea*—and probably plenty of childhood nightmares—does feature a villain-killing octopus.)

In general, octopuses have a distinct taste for shellfish. Giant Pacific octopuses have been known to feast on gooey geoduck and unsuspecting gulls, and have even been seen digging up clams for dinner. Smaller species of octopus will snatch up lighter fare, such as hermit crabs. Dungeness crabs and even lobster have been found in octopus stomachs. Janet Voight of the Field Museum discovered deep-sea octopuses with guts full of hydrothermal vent gastropods and polychaete worms. She has watched some of these animals in action via the U.S. Navy–owned deep submergence vehicle *Alvin*. Their arms were buried under the substrate, where, she figured, they were probably digging for worms and other buried animals. Above all, octopuses are hunters rather than scavengers like some of their benthic brethren, such as lobsters. They prefer live prey, be it a snail or a shark (so perhaps we shouldn't feel so bad about any affinity for the Korean live octopus dish).

Octopus stomachs do not have the right enzymes to digest plants, so seaweed is off the list. Likewise, the lucky beasts can't digest fat; it just passes right through them (to be snatched up by feces-feasting fishes that swim behind them, yum).

Octopuses eat about a third of their meals while they are out and about. Biologist Jennifer Mather found that common octopuses she was studying brought the rest back to eat in the comfort and safety of their own dens. Some have even been seen saving leftovers—perhaps a crab leg or two—for later.

As Aristotle noted in his *History of Animals*, "The octopus for the most part gathers shellfish, extracts the flesh, and feeds on that; in fact, fishermen recognize their holes by the number of shells lying about." It "is neat and thrifty in its habits: that is, it lays up stores in its nest, and, after eating all that is eatable, it ejects the shells and sheaths of crabs and shell-fish, and the skeletons of little fishes."

The octopus isn't *always* so neat as to take out its refuse regularly. Which is fortunate for folks like Roland Anderson, who was able to study the octopus diet by collecting discards of our own: beer bottles. "Little red octopuses in Puget Sound like to live in beer bottles," he says. If you come upon a bottle an octopus had been living in, "you could tell what the octopus was eating by examining and identifying the shells," he explains. Through that study he also got a little lesson in octopus architectural preferences. "I determined that the octopuses like brown beer bottles better than green beer bottles or clear beer bottles, and they like stubby beer bottles better than long-necked beer bottles," he says. Perhaps they like the shading for extra privacy and the wide space for a roomier living area. So if you're determined to litter at sea but still want to keep the little octopuses in mind, I suppose Red Stripe bottles might be the way to go.

All-terrain Killers

I took my failure to find a wild octopus in Puerto Rico as evidence of their masterful camouflage capabilities. As Aristotle observed, this tactic is not just for avoiding becoming dinner. "It seeks its prey by so changing its colour as to render it like the colour of the stones adjacent to it."

These hunters can be superstealthy, as biologist James Wood learned the hard way with his very first octopus. Then a high schooler, he had already been keeping aquariums in his bedroom for years, but these increasingly complex marine systems had never held anything so clever as an octopus.

On the last day of summer vacation, on a trip to the Florida Keys, Wood caught his first pet octopus. "The next day was the first day of school for that year," he recalls. He had read about octopuses' escape-artist skills, "so I just figured, 'I'll just go for overkill, and then tomorrow I can deal with it,'" he says. So he duct-tapes the lid onto the tank and then weighs that down with a four-by-four piece of wood. In the morning, however, the octopus is nowhere to be seen. The lid is still on, and the tank itself isn't so big or elaborately decorated that the animal would have gotten lost inside of it. An escaped, dead, dried-up octopus would be a good excuse for his parents to put an end to his in-room aquarium collection. So he starts frantically tearing his room apart. "I'm just looking under everything, and I can't find it," he recalls. By that point, he says, "I'm late for school. Everybody else is probably worried about making good impressions, what clothes to wear. I'm lucky if I have my underwear inside of my pants." That afternoon he looks again—and still no octopus. So he removes everything from his room. Nada. So he starts searching the rest of the house—in corners, around the bathtub. "I still can't find her," he says. Three days pass, and still no octopus—dead or alive. It is only then that Wood gets the bright idea of looking at the tank from below. So he gets his scuba diving light out and peers up.

And there it is, under the bottom filter. The gravel above appeared not to have been disturbed. He still has no idea how it got down there. But he got to keep it. And the octopus quickly learned that it no longer needed to hide—for hunting purposes or for its own safety. Being the courier of food, Wood became its favorite person. After just a couple of weeks of keeping the octopus, when Wood walked into the room, it would come out of the rocks and move around at the front of the tank. "If this had been a vertebrate," he says, "we would call that behavior begging."

But octopuses in the real world, of course, are not going to get such easy handouts.

The octopus actually has quite a few methods for grabbing dinner. One is to lie in wait, in a crevasse or camouflaged against the bottom, until a tasty-looking treat wanders by. Another more proactive shopping strategy is to scour nearby surfaces for food and keep tasty tidbits safe under its web. This combing method has been nicknamed the "webover." To pull off this trick, it spreads the web that connects the

An *O. briareus* spreads its web wide.
(Denise Whatley/TONMO)

upper part of its arms over a surface and can feel underneath it with arm tips or suckers for small crabs or other snacks. Meanwhile, other arms can search the surrounding area for more morsels, which it can stash under the web for safekeeping—and eating.

Thanks in part to its good spatial memory, an octopus seems to know when it has covered a territory and it might be time to move on to new hunting grounds. Once it's exhausted the areas within easy reach of one den, whether it's taken days or weeks, the hungry octopus will move on to a new neighborhood.

There are plenty of tales about octopuses in public aquariums that venture out of their tanks at night to eat fish or crabs in a neighboring tank and then return to their own enclosure before daybreak. But these are the lucky ones. As a general rule, "if they do climb out of their tank, it's unfortunate," the National Zoo's Tamie DeWitt says. "Very rarely do they get back *into* their own tank." More often than not, an escaped octopus will find itself on the floor, and even if it knew to return to its tank, it would likely not be able to climb back up and in. "That's always a problem, because you come in in the morning, and the octopus is dead." (They also often commit unwitting suicide by pulling a plug out of their tank and draining it. Such are the hazards of curiosity augmented with extreme dexterity and decent strength.) Even in one seemingly secure tank at another zoo, an octopus moved a brick to escape through a PVC pipe that was just an inch in diameter and way above the water level. That PVC pipe, unfortunately, led to a draining system, so, DeWitt says, "they found the octopus in the morning" (read: not in good shape; dead, that is). As many scientists will explain, we don't really know exactly how long octopuses can survive outside of the water because crafting such an experiment would lead to lots of dead octopuses (and little more than a number to satisfy our curiosity, rather than a step toward more substantive research). They have been filmed out of the water for at least several minutes, but their full capacity likely depends on ambient humidity and temperature.

Shallow-water octopuses occasionally come ashore in search of food. "The octopus is the only mollusk that ventures onto dry land; it walks by preference on rough ground," Aristotle noted. It's not a pretty sight—all of their underwater grace gone like last night's dream of flight. Jean Painlevé's two octopus films both feature this unsightly lurching, which makes the octopus seem all the more alien. *The Love Life of the Octopus* (1965) even starts with a voiceover of one of these scenes: "Octopus . . . cephalopod . . . horrific creature . . . flabby, without a shell. It slithers at low tide." Strange 1960s sci-fi music (similar to eerie tracks in the 1976 campy horror film *Octaman*) plays in the background. More recent footage, taken as a home movie at the Fitzgerald Marine Reserve preserve in California, shows an octopus pulling itself, painstakingly, among kelp plants to the amazement and amusement of onlookers.

Although that California octopus seemed a bit directionless, "they often know which way to go," biologist Lydia Mäthger says of land-crawling octopuses. Despite likely having limited vision in the open air—and a very low vantage point from which to see the way—they usually crawl in the right direction toward a new tide pool or back into the ocean. Mäthger suggests that maybe they can sense fine moisture differences with their outstretched front arms, or perhaps they just follow gravity and tend to go downhill, where there is more likely to be water gathered.

These itinerant hunters will often go to great lengths to set out for new territory. As much as octopuses might enjoy a cozy hiding spot for a while, when they decide it's time to move on, often little can stop them. Wood found this out the hard way. One day at the beach he captured an octopus he wanted to keep—the only problem was that he had two big buckets but no lids. So, knowing their tendency to escape, he put the octopus in the bottom bucket with some water, and then he filled the top bucket about two thirds full of water and put it on top of the first bucket—a weight he estimates to have been about thirty-two pounds. Wood surveyed the seam between the two buckets and could see only about a millimeter-wide crack. And the octopus's hard beak would have been bigger than that sliver of an opening. His friends thought he was going a bit overboard, but he could tell he'd landed a "really scrappy" one, as most of its arms showed signs of having been chomped on at one point or another. With the plucky octopus stuck safely under thirty-two pounds of water and little more than a millimeter crack to the outside world, Wood and his friends headed out to hunt for more specimens. "We come back," he recounts, and the buckets are still stacked, but "there is no octopus. It has gone." To this day, the only explanation Wood has is that the octopus must have lifted the heavy water-filled bucket up above it in order to sneak out of the lower one. "That's what had to have happened, but how that happened, I have no idea." His amazement helped ease the sting of having to go home sans octopus.

Advanced Vision

Even in tanks—but especially in the wild—the world a wild octopus inhabits might as well be another planet. Sound and light travel differently than they do in the air. Pressure and gravity are altered. A predator can come from absolutely any direction. To prevail in their pelagic universe, the octopus has neglected some senses—such as hearing—while ramping up others—such as taste. And it has made some amazing advances in the field of vision that we're still trying to parse.

Looking an octopus in the eye is like looking through more than half a billion years of evolution, twice—back in time down our path to an ancient common ancestor and forward again through the octopus's. As if it wasn't already strange enough to stare into the eye of a seemingly sentient cephalopod, you're encountering two stunning examples (the octopus's and ours) of Nature's development of a sharp hunter's vision with sophisticated lens-based focusing. Whatever genes lay in our last common ancestor—likely a sightless worm that probably had little more than some light-sensing cells on its body—they managed to morph into two versions of nearly the same organ.

We've both developed lenses, retinas, corneas, and irises, resulting in "camera" eyes. And some 69 percent of genes examined in one study were expressed in both octopus and human eyes. Biologist Andrew Packard proposed that it was eons of competition with vertebrates that led their eyes to be so similar to ours.

The octopus even has one up on us physiologically, in addition to its being able to detect polarized light. Our optic nerve is attached to the back of our eye, which is fine, except that it creates the small blind spot in the middle of our retina. The octopus's optic nerve goes around the outside of the retina, leaving it with uninterrupted vision in the whole eye.

To help it as both a predator and a prey animal that lives in a fully three-dimensional world, the octopus has especially wide eyes—all the better to see in many directions. The position of its eyes helps it to get a much more panoramic perspective. The horizontal pupil is also rotatable, so no matter which way the octopus is oriented, it is cued in by the statocysts and stays horizontal. To John Steinbeck, these eyes seemed downright demonic, as he noted in his novel *Cannery Row*: an octopus waits for unsuspecting prey "while its evil goat eyes watch coldly."

To many others, however, the eyes have been a great source of inspiration. Albert Titus, an engineer at the University of Buffalo, has used the octopus retina as a model to build an artificial visual processing system. With the "o-retina," which is made from silicon chips, he and his team hope to diversify the type of visual information that can be gleaned from robotic explorers in exotic environments—whether in the deep ocean or deep in space.

The octoeye has also lent some insight on how to improve images we take with our cumbersome curved-lens cameras. In research funded by the military, scientists at Case Western Reserve University have honed the cheekily named GRIN (gradient-index) lens. The thin, plastic layers in the lens not only get rid of the bulky traditional camera lens, but they also allow for clearer imaging—especially around the edges.

And yet the octopus's own visual acuity has been a matter of some debate.

Recent research suggests that octopuses are actually rather nearsighted. They might not be able to see too clearly beyond eight feet away, lending accuracy to the smart but myopic Marvel comic character Doctor Octopus.

As far as scientists have been able to discern, octopuses can't see color. "There has been a lot of work done on the eyes, and there is no known mechanism of color perception built into that eye," Roger Hanlon says. "Maybe some smart person will find something else in there, but it will be totally new." And that dead end has been what has led them to look at the skin for visual sensors—"somewhat in desperation," Hanlon confesses—where they landed on the light-sensing opsins.

With what we know about their biology now, however, Lydia Mäthger says, "everything is going to appear to them a shade of whatever color they could see best," which, after examining the makeup of their eyes, researchers determined is green. So "they basically see everything in shades of green," she says. But, of course, if that's all they know, it will just be their version of gray.

And in much of the ocean, color vision isn't as relevant as it is on land. "The deeper you go into the sea, the fewer colors there are," she says. This helps to explain why deep-sea octopuses are pale white or deep red (which appears black deep down) and don't bother with ink or fancy color changing.

There are, of course, plenty of shallow-water animals that have gone nuts with their color vision. Mäthger recommends Googling "mantis shrimp color vision" next time you're online. "It's insanity," she says. Instead of the three main "color" wavelengths we can see, they have twelve color vision channels, which gives them exponentially more colors in their rainbow. "Plus polarization," she says. "It's just insanity! And that's a bloody shrimp. And they share the same kind of habitat. So to me, it's a little astounding that cephalopods just wouldn't have color vision," she says. And, she adds, some squid have been shown to have color vision, so it's possible that there *are* some octopus species out there that have this ability, but they just haven't been one of the few species that have been tested extensively in the lab. "We have a tendency to generalize things," Mäthger says. And with the finicky octopus, our sample size has been relatively small.

To augment their sight, like many other aquatic animals, octopuses have developed the ability to see polarized light. The light that we perceive is generally a wide, messy array of waves coming in from many directions, including hazy reflected light, say, from a sun glare on a body of water. Polarized light, however, is confined to waves that come in only at a single angle. We've been able to create polarized sunglasses, which can streamline light and get rid of the chaotic reflected light.

Hanlon has not completely given up on looking for other ways an octopus might see in more than just one color. While filming one camouflaging octopus in dim light conditions some seventy feet below the surface, Hanlon switched on and off his bright light, changing the visual conditions. And with each light change, the octopus had "a precise color match" despite the general periods of darkness, he says. That "makes me believe that they ain't color blind. Some active process was going on there."

Full-color or gray-scale, octopus vision is definitely not slicing the world into a prismlike landscape, as some bugs might, a detail that was apparently unknown to the creators of the film *Octaman*. When the audience is treated to a creature-eye-view of the world, it's split into segments as if from a fly's perspective. (Of course, while we're at it, the Octaman's deadly assault style is also a little questionable. The upright, zombielike creature, whose extra arms are tied to the actor's own appendages, fells foes by whapping them with its arms. This leaves the victims bloody and maimed or bloody and dead. If any cephalopod is going to do much damage with just one swipe, it's much more likely to be a squid, some of which have sharp hooks on their extremities.)

The octopus might rely primarily on contrast and movement to detect dinner, just as we tend to look over if we notice something moving out of the corner of our eye. "Movement detection is what brings them their lunch," Mäthger says.

However they are seeing the world, octopuses likely do rely on vision for a large part of their perception, based on how much brain power is devoted to the task. Like us, the octopus has large optical lobes. They seem to at least be able to see well enough through water, glass, and air to spot their favorite—or least favorite—researcher approaching to know whether to get ready to greet them at the front of the tank or squirt them with water.

Ocean of Noise

Although the octopus's visual capabilities remain keenly studied, researchers only recently settled the debate as to whether octopuses and squid are endowed with the sense of hearing. Most of the ocean's vertebrate animals, such as fish and marine mammals, possess rather complex anatomies for picking up sound waves. But because the octopus lacks these features—an adaptation that helps it survive immense pressure and squeeze into tight spots—scientists weren't sure how it might be able to make out any audible vibrations.

Hong Yong Yan, a physiologist at the Taiwan National Academy of Science, and his colleagues, however, finally determined that both squid and octopuses use tiny hairs in their statocysts, the internal structure that also helps them orient themselves, to pick up on a narrow range of ocean sounds. Benthic octopuses seem to be harder of hearing, however, than their open-ocean dwelling cephalopodian cousins, the squid. Common octopuses can pick up sounds ranging only from 400 to 1,000 hertz (oscillations per second)—we humans, by comparison, can pick up sounds between 20 and 20,000 hertz. Yan suggested that the octopus's limited hearing might be because the higher frequency sounds don't travel more than a few feet in seawater, making it even less likely that they'll make it far over an uneven ocean floor to the "ear" of an octopus.

And purists point out that "hearing" isn't even quite what we should be calling this sense of theirs. What a cephalopod does underwater is "detect vibrations." Other research has suggested that some cephalopods also possess specialized cells on their arms and mantle that can detect vibrations between 75 and 100 hertz. Much as a moth has evolved to be able to pick out a bat's high-frequency sonar sounds—in order to avoid becoming food—octopuses, squid, and cuttlefish might be able to sense the dinner-detection frequencies of some of their potential predators, such as whales or dolphins. These sound-detecting systems also might help cephalopods detect dinner—even in the dark, biologist Ángel Guerra notes.

Even with their limited hearing, octopuses, like dolphins and whales, are vulnerable to loud noises underwater, possibly including those associated with shipping and offshore drilling. I had wondered about this while visiting Vigo, which has turned into a major shipping hub. Michel André, of the Technical University of Catalonia in Barcelona, and colleagues found that octopuses—in addition to squid and cuttlefish—ended up with damaged statocysts when exposed to low-frequency sound (between 50 and 400 hertz), even if it wasn't for very long. With busted statocysts, octopuses might have a harder time hunting and avoiding predators, according to André.

Octopuses are typically the quiet, retiring type. But scientists are finding not only that can they hear, but also that they might be able to make noise. Guerra and his colleagues documented one octopus that made a sound "like a gunshot." The blast of sound was also accompanied by "distinct lights," they noted in a 2007 paper about the surprising observation.

"It's a new noise in the sea," Guerra says. New to us at least. In their paper, he and his colleagues suggest it might "be a defensive strategy to escape from vibration-sensitive predators." The octopus in question, a common octopus, was being filmed in the Mediterranean while it was being attacked by a grouper. As the researcher approached, the octopus was able to escape the fish. But the researcher followed the octopus as it fled the scene. "It went down toward the bottom and released ink before hiding underneath *Posidonia oceanica* plants," the researchers wrote in their paper describing the incident. While it was hiding, the octopus flashed red and ejected more ink; "simultaneously, a strident shotlike sound of 135 milliseconds' duration was produced by the animal." The frequency pushed the needle up to 24 KHz. The source was likely an extremely high pressure jet of water shot out from its funnel that created a severe pressure differential so that microbubbles in the seawater expanded and then rapidly collapsed. These bubbles would also have changed the reflection of the sunlight, producing the mysterious flash of light.

Squid have occasionally been heard producing a faint popping noise, but a bang of this magnitude is closer to the cavitation feats achieved by some species of shrimp. Whether the octopus's burst of noise was a specific defense tactic that it can call upon at will or this observation was a fluke remains to be seen—and heard.

Perceptive Limbs

For capturing its lunch, the octopus's primary weapon—other than, arguably, its intelligence—is its arsenal of arms. Actual octopus arms likely aren't able to lift three tons, as each of *Spider-Man*'s Doctor Octopus's arms reportedly can. So don't presume they can take down mega sharks—unless the octopus in question truly is gigantic, as in the 2009 B movie *Mega Shark versus Giant Octopus*, starring Lorenzo Lamas, which features these two multiton prehistoric beasts unlocked from their frozen 18-million-year-old battle when a chunk of glacier melts.*

But it's not just the sheer arm strength that is involved in catching dinner. In the wild, the octopus often relies on its dexterous arms to go rooting around in crevasses for crabs or other crawling foodstuff. These arms can feel and even taste their way to dinner and then react to capture it. With all of this limb autonomy, octopuses were, for a long time, presumed to be blind (so to speak) about where their arms were and what each one was doing while it felt along for food.

Certainly many small operations are likely being controlled from within the arms themselves—or at least below the brain. But there must be some sort of central perception and processing going on, or else the animal would just be a big eight-armed pile of mess. "If you look at octopus behavior, it does not make a lot of sense" that it would have no

*This reanimated battle costs several submarines and one of the octopus's arms (which seems to get bitten off twice after some economical film editing). But finally, with the shark struggling in the octopus's grip, they both sink down into the dimness, although it would seem that the megashark wins, because its character resurfaces in 2010 to do battle once more in *Mega Shark versus Crocosaurus*.

idea where any one of its arms is at any given time, Hebrew University's Michael Kuba says. "Anybody who uses common sense should see that an octopus must know where its arm is because otherwise it's very hard to function," he says with a laugh. But "it's one of those problems science sometimes has" in which researchers don't always use common sense in developing hypotheses, he notes. Somehow, "the brain *has* to decide which arm to activate, at what speed," Kuba's colleague Binyamin Hochner explains. How does it know this? Partly as we do, probably, with an internal little homunculus (theirs would have to be an octonculus) as a mental schema for their bodies. And like us, the octopus is probably also using some visual intel to control its limbs.

Our own visual sense of self is so ingrained that it seems hard to believe it doesn't have deep evolutionary roots. Recent human experiments have shown that if you create the optical illusion for a person of having a third arm, and if it's in a believable location, a person's brain will start to perceive it as part of his or her body plan. The person will even react when the false arm is threatened with a knife. Hochner, Kuba, and their colleagues designed a slightly less violent experiment to test the octopus's ability to visually assess its arms. They drew upon the octopus's natural behavior of sending individual arms into holes to look for food. They built a clear plexiglass maze that had one central vertical tube and three separate compartments just beyond it. They first trained the octopus that food would be waiting in a compartment with a black dot on it. For the experiment, the octopus—upon seeing which compartment had a black dot on it—would need to send one of its arms through one of the tubes, through an air gap (so that it couldn't receive any food smells through the water), and into the compartment where the food was. The researchers randomly changed the location of the dot and food ten times for each animal each day for up to three weeks. Each assay, an octopus got only three minutes to complete the task—and no second guesses. Six out of the seven common octopuses passed the test by getting the right box five times in a row, the researchers described in their 2011 paper.

Watching footage of the experiment, Hochner and his team noticed that when the animals couldn't see the box with the dot—because they had positioned themselves poorly—they weren't likely to select it. "Animals learned to orient themselves to get an unobstructed view of the target," they noted in their paper. When researchers tried the same experiment with an opaque maze, the octopuses got the answer right only randomly. This suggests not only that octopuses can figure out the location of one of their arms based on visual information but also that they can visually guide it to a target. In the real world, this would mean that being able to spot a tasty crab would help them accurately capture it with an arm. So even though octopuses are impressively skilled at hunting "blind," aided by tactile and taste/smell senses locally on the arms, being able to help guide their body parts centrally with sight makes them even better predators—and improves general coordination to help keep themselves from becoming prey.

In addition to keen central visual control, octopus arms also have excellent taste (their sense of it as well, certainly, as when served in a nice salad). As beings that evolved isolated in our lonely air pockets, it is hard for us to imagine what it means to be physically engulfed in a much more conductive medium like water. Sure we have smell, but the dense world of an octopus is literally swimming in chemical signals. (This might even be how lonesome octopuses detect a potential mate.)

But octopuses do not just passively receive smells and tastes coming in through the water currents. Because their suckers are equipped with chemoreceptors, they can taste what they are handling. And a sizable portion of their brain appears to be devoted to deciphering this information. So instead of waiting until you put that peanut butter and jelly sandwich in your mouth, if you were an octopus (aside from having to deal with a soggy sandwich), you would be tasting it as soon as you laid hands—er, suckers—on it.

This physiological feat was rendered artistically (whether intentionally or not) by Timothy Hawkinson in an artwork commissioned by the Getty Museum. The 2006 piece, simply titled *Octopus,* features a collaged octopus, its underside out, spanning the canvas. Upon closer look, you notice that instead of suckers, the artist has painstakingly placed images of different individual human mouths. Unsettling—but surprisingly accurate.

Beyond a refined sense of taste, octopuses' tactile sense is also highly developed. This is key, as Mäthger points out, because they do often have to grab their prey without spotting it first. "Some of these guys poke an arm into the holes just to see what's in there," she notes. With each sucker and inch of arm being so sensitive, octopuses rely on their local reaction and control instead of having to bother the main brain with a tickle here and a bump there.

Deadly Feast

Once an octopus at last has its quarry in its grasp, it now is faced with the challenge of eating the unfortunate victim. In the absence of metal claw crackers, tiny two-pronged forks, and plastic lobster bibs, the octopus has devised its own way of consuming its mostly hard-shelled meals. It can use its suckers and strong arms to pull open many two-part shells and tear the animal out. If that doesn't work, the octopus can use its beak to chomp into its food (or a pestering human handler). If this proves too tough still, it can drill into a shell with its sharp beak, serrated radula, and sharp salivary papilla. If this is *still* unsuccessful, an octopus can secrete an acid to dissolve the hard shell's calcium carbonate.

To extract the victim, an octopus can eject venom to paralyze the animal with an enzyme that also breaks down some of the muscle attaching it to its shell. The venom enters the octopus's saliva from special glands located just behind the brain. And, amazingly, many octopuses seem to have figured out the best sites (a weak area in the shell or a place where a muscle is attached) for injecting their deadly doses of poison.

Although almost all octopuses are thought to have some venom, most are of little threat to you or me. But working with the supervenomous ones can be quite dangerous. As Mäthger notes, when they were studying the blue-ringed octopus, they even made sure they never worked with the animals while they were alone. The bites are but the smallest of punctures and, apparently, often don't feel like much at all.

But if the animal injects some of its venom, called maculotoxin, its victim receives a frequently fatal dose of tetrodotoxin (also known as TTX). Tetrodotoxin is a neurotoxin that stops sodium ions from moving across neural membranes. This same toxin has also been found in some species of crabs, fish, newts, sea stars, snails, and worms.

The poison isn't coming from the animals themselves but rather from bacteria that live in the animals. For the blue-ringed octopus, this tetrodotoxin-making bacteria has found a cozy home in its glands, and the little octopus is lucky to have it there. A bite from the blue-ringed octopus can render an adult human paralyzed and unable to breathe within ten minutes. One study found that an adult blue-ringed octopus that weighed less than an ounce held enough of its supertoxic venom to paralyze and kill ten full-grown people. No antidote is currently known. Pretty serious for such a small little sucker.

This deadly strategy is not just a matter of instinct but also one of chemistry and of evolution. Octopus venoms seem to have evolved to target specific prey. "Over millions of years of evolution, venoms have, in fact, decoded the nervous system" of other species, Brooklyn College's Frank Grasso says. These varieties of venom are now being investigated for compounds that could have biomedical applications for humans. It might sound like a strange or even dangerous lead, but researchers have already probed snail venoms for painkillers. Studying extreme octopuses has also yielded some surprising scientific results. One expedition to the Antarctic, led by Brian Fry, of the University of Melbourne, uncovered an octopus venom that still works at freezing temperatures.*

How an octopus disarmed, dismembered, and consumed its prey remained a mystery for ages. With a quick swoop, crabs or other morsels are swept into the webby nether region and out of sight. Sometime later, an empty shell will emerge. But the steps in between are hidden from human view.

*While surveying those deep, cold waters, Fry and his team also discovered two entirely new toxins via octopus venoms, as well as a new species of octopus altogether (which received the unglamorous, temporary name of *Parledone species 1*).

The octopus web has fascinated and horrified onlookers for ages. This obscuring "membranous umbrella," as the nineteenth-century naturalist Henry Lee described it in *Land and Water* (republished in 1873 in the *New York Times*), was particularly vexing to him. He professed that he "wished to be underneath that umbrella with the crab, or (which was decidedly preferable) to be able to see what happened beneath it without getting wet." To get a glimpse, he dangled a crab on a string by the glass side of the tank, and the resident octopus took the bait. He was at last rewarded with a clear, if chilling, view:

> In a second the crab was completely pinioned. Not a movement, not a struggle, was visible or possible—each leg, each claw, was grasped all over by suckers—enfolded in them—stretched out to its full extent by them. The back of the carapace was covered all over with the tenacious vacuum-discs, brought together by the adaptable contraction of the limb, and ranged in close order, shoulder to shoulder, touching each other; while between others, which dragged the abdominal plates toward the mouth of the black tip of the bard, horny beak was seen for a single instant protruding from the circular orifice in the center of the radiation of the arms, and the next had crunched through the shell, and was buried deep in the flesh of the victim.

Lee likened the action to the commonplace sight of a mouse being pounced upon by a cat or a minnow being snatched up by a perch. But nevertheless, he acknowledged some uneasiness about the whole ordeal he witnessed: "There is a repulsiveness about the form, color, and attitudes of the octopus which invests it with a kind of tragic horror."

Steinbeck, too, seemed captivated and appalled by this hidden violence, as he describes in *Cannery Row*: "It leaps savagely on the crab, there is a puff of black fluid, and the struggling mass is obscured in the sepia cloud while the octopus murders the crab."

I get to see this mysterious process firsthand when I meet Pandora, the still-small new giant Pacific octopus at the Smithsonian National Zoo in Washington, D.C. It's a warm, muggy May morning, and I have only a couple of hours to spare before I have to catch a train home to New York. I wind my way to the back of the zoo, past the pandas, elephants, and great apes, to where the invertebrates live. Tucked away humbly out of the main thoroughfare, the invertebrates live in a modest building behind the reptiles—out of sight, much as they are in our daily consciousness.

Inside the exhibit, it is hot, humid, and dim. I round a few corners, and there she is. Pandora is smeared to the top corner of the glass, as if awaiting my arrival. Her eyes are in a half-asleep, bored squint as her mantle hangs just at the waterline. The fine tips of her arms dance around in the water currents, but other than that, she remains a stationary sentry.

In the cephalopod room, she's not getting nearly as much attention as the large, gracefully suspended chambered nautiluses or the cuttlefish, which put on a show of flashing arms and tentacles as they get fed by a volunteer. No, Pandora just sits quietly in her corner as kids proudly point out "Octopus!" and move on.

By happy coincidence, the day I visit is the Invertebrate Exhibit's twenty-fifth anniversary, which the staff and volunteers are celebrating with festivities and an official cake cutting. I spot a blond woman dancing with a couple of young kids—it's the always-in-motion Tamie DeWitt. She tears herself away from the party, and we head over to the exhibit hall's prep area to grab some tasty crab morsels from the refrigerator. She stuffs them inside a red whiffle ball and asks me if I'm ready to get my hands wet.

The top of Pandora's tank is sealed off with slats of clear plastic that lock together from the top—very secure. "You always have to close the lid and latch it because the octopus is watching everything you're doing," she says. We both look around the tank for Pandora. It takes a minute to spot her; she's now hunkered down in a back corner. DeWitt tells me to splash the ball around a little bit to get the octopus's attention.

And sure enough, it works. Pandora starts to slink an unbelievably long arm and then her body over. DeWitt tells me to pull the ball away and let Pandora grab onto it so I can slowly pull her into the middle of the tank. Pandora eventually obliges, and she starts to wrap the slender tip of her arm around my finger. Sucker by muscular sucker, I can feel the pure strength of this animal. I get a flash of insight into the terror they can evoke in anyone at sea who fears being overpowered and dragged down to a watery doom. DeWitt had told me how strong even these acclimated octopuses can be. "You're never going to win a tug of war with an octopus—especially a full-grown octopus," she says. "Those arms of an octopus are pure muscle"—not to mention "the power of the sucker," she notes, warning me about the hickey threat. And, she had reminded me, the beak is just like a bird's and "very, very powerful." It could easily break the skin, even if the animal is only being curious.

But this placid kraken wants only the crabmeat-filled ball, which I'm sure she can sense with the chemosensors on her suckers. Pandora wraps an arm onto the ball, and DeWitt tells me to let go (I reluctantly obey, not wanting my brief octopus contact to end). At the zoo they try not to handle the octopus, because it quickly winds its arms all over human arms, and then the process of tearing them off can result in torn octopus skin. Their cuts can easily get infected, "so it's really for the safety of the octopus that we don't do a lot of hands-on," DeWitt had warned me by phone before my visit.

Pandora quickly whisks the ball underneath her web and plants herself against the back wall backdrop. You can see the bump where the ball is, but it's nearly impossible to see what she's doing with it or even to detect much movement. Pity the live catch that must undergo such terror in the wild.

At last, perhaps as a send-off (or perhaps only because she had dislodged the crabmeat from the ball and wanted to let us know), Pandora makes an elegant push off the side of the tank and floats upside down, suckers to the surface with the red ball still stuck to her mouth. It's a

Pandora, the National Zoo's giant Pacific octopus, displays her empty ball. (Katherine Harmon Courage)

strange and wonderful sight to behold before I have to run to catch my train.

But oddly, all of these perceptive pursuits, cunning catches, and ferocious feasting are for an animal that lives but a short, lonely life. The octopus, in fact, has one simple goal in its brief time in the ocean: to survive long enough to meet one mate before dying.

CHAPTER EIGHT

Sex and Death

The octopus, as many researchers are fond of saying, lives fast and dies young. Despite all of the bodybuilding, color changing, and garden tending, the octopus leads but a brief life of quiet solitude.

Some deep-sea octopuses might live longer than the giant Pacific octopus's few years, but research has been scant. One octopus off the coast of Peru lives far below the surface in the dark, chilly seas and lays enormous eggs. It might live for as long as nine years, some researchers hazard. But some small species live out a full life cycle in less than six months.

"These animals have marvelous vision systems; they have this marvelous camouflage; and they have these huge brains," biologist Roger Hanlon says. "All of that packed into a short life is really weird."

Nevertheless, they seem to be in a hurry to grow up, settle to the bottom, mate, and fade away. And why not, the Field Museum's Janet Voight points out. If you, as an individual, spent extra time bulking up while everyone else was mating, you might have lost your chance at pairing off—no matter how impressive your muscles. So for these animals, it's a race to the finish. And that strategy seems to work pretty well. After all, octopuses, squid, and cuttlefish have been doing this for hundreds of millions of years, and they seem to be pretty successful so far.

To have had so many millions of generations, you've got to have a pretty good procreation system. Some mollusk cousins, including oysters, are hermaphroditic and can change sexes based on mating needs. The octopus is decidedly single sexed. And if you're wondering how to tell a lady octopus from a male one, Aristotle has a few tips:

> The males have a duct in under the oesophagus, extending from the mantle-cavity to the lower portion of the sac, and there is an organ to which it attaches, resembling a breast; in the female there are two of these organs, situated higher up; with both sexes there are underneath these organs certain red formations.

Or you could take the old octopus-walks-into-a-bar joke as guidance. If you haven't heard it, it goes a little something like this: An octopus (or a guy with an octopus) walks into a bar. The octopus, apparently, is the best musician in the county. So the animal is challenged with a guitar, which it plays expertly. Piano? No problem. Finally someone gives it a set of bagpipes. The octopus fumbles with it for a moment. The audience starts to boo. "Aren't you going to play them?" someone yells. "Play them?" The octopus retorts. "I'm going to take her upstairs as soon as I can figure out how to get her pajamas off."

Octosutra

Scientists are still learning how these lone animals find each other in the great big ocean when it does come time to mate—with no help from social groups or missed-connections Web sites. There often aren't even hard-and-fast mating seasons. Spawning seems to happen pretty much year-round for the common octopus—with a couple of peaks in the spring and autumn.

One way these lonely creatures find each other might be by scent. Just as moths use pheromones that float through the air, octopuses and other cephalopods might send out chemical cues when they're mature and ready to mate. They can pick up this information from the chemosensors in their suckers and from the area around their mouths. Millersville's Jean Boal has found that octopuses can "detect the differences in odor between males and females," she says. And, curiously, "they do detect the odors of eggs. And why? What for? I don't know."

Once a potential mate *is* found—usually it seems to be the male seeking out the female—a courtship color display or even a chase is usually in order to get the job done. If multiple males happen upon the same female who is ready to mate, they will often fight over the chance to bed her. Lady octopuses can, however, accept sperm from more than one male.

Some species engage in a brief, close physical union. Fittingly, for Bond aficionados, the blue-ringed octopus (of *Octopussy* fame) engages in an especially lusty encounter, with the male grabbing the female's mantle with his arms and repeatedly inserting his hectocotylus into her siphon for minutes on end. For the blue-ringed species *Hapalochlaena lunulata*, the coitus often continues until the female physically removes the male. The little-studied larger Pacific striped octopus is perhaps the most social mater; it apparently does the deed face-to-face, sucker-to-sucker, as Aristotle once imagined all cephalopods did:

> Molluscs, such as the octopus, the sepia, and the calamary, have sexual intercourse all in the same way; that is to say, they unite at the mouth, by an interlacing of their tentacles. When, then, the octopus rests its so-called head against the ground and spreads abroad its tentacles, the other sex fits into the outspreading of these tentacles, and the two sexes then bring their suckers into mutual connexion.

But for other species, sex is a rather removed affair. Rather than mounting the female directly, the males of these species will sidle up and hand over the packet of their sperm—the spermatophore—which slides down a groove in its specialized hectocotylus arm and into the female's mantle. These spermatophores look a little like thin bean sprouts in some species, and are surprisingly big for such little animals. (Janet Voight showed me a vial of preserved spermatophores in her office at the Field Museum, and they were at least an inch long.) At the tip of this multitalented arm is a spoon-shaped structure called a ligula, and in some species, it can become erect to prepare for mating. The male's hectocotylus, as Aristotle described it, is not a gentle organ. "The last of his feelers he employs in the act of copulation; and this last one, by the way, is extremely sharp." This, he notes, "enables it to enter the nostril or funnel of the female."

Indeed, octopuses must get their lovin' however they can. And as the narrator in Painlevé's *The Love Life of the Octopus* (1965) notes, "There's no officially sanctioned position for doing that." On screen, the short film shows a male octopus, a pale white color, that has managed to get his hectocotylus into a female's opposite funnel by reaching his arm across her face. At first she resists, pulling him by the ever-stretching arm, but then he manages to rope her in, like a lassoed animal, and bring her closer so he can finish the deed.

Christine Huffard, a biologist with Conservation International, spent hours—and sometimes days—scanning the reefs around Sulawesi, Indonesia, following the quiet lives of some 167 of her small subjects, *Abdopus aculeatus*. She saw that males could practically mate with the girl next door without leaving home. One could simply extend his hectocotylus out his den and reach it into a nearby female's den. Although it might seem cold to us, this mating strategy could be a life saver. It allows these nearby octopuses to procreate without risking a big public display of affection, which could leave them vulnerable to passing predators. It would also explain why she saw males and females living in quite close quarters.

Although these mated males would chase away other male competitors—especially in this particular area, where males outnumbered females by almost two to one—these pairs weren't exactly monogamous. If the male wandered off in search of food, another male might stop by to mate with the female. And some male octopuses have a different tactic for getting goods to a female. Rather than keeping a lady around, this male is a "sneaker." He uses his camouflage to creep up on a defended dame—even hiding his mating arm's markings—to slip his spermatophores to the female. Huffard observed this approach pay off a few times. But it wasn't foolproof. On one occasion, a defending male did see the sneaker. But because this second male had disguised his hectocotylus, the home-field male thought it was his lucky day: another female had arrived! But when he began making romantic overtures in the would-be cuckolding male's direction, the sex confusion became clear, and he set upon the visiting male more violently to chase him off.

Not all octopuses are lucky enough to find a mate at all. But the Seattle Aquarium gives these lonesome animals at least a chance at love. On Valentine's Day, the aquarium's biologists allow two captive adult octopuses to mingle for a brief opportunity to breed before they release them into the nearby sound. In 2012, they introduced the three-year-old female, Mayhem, to Rocky, the male. Not exactly a little lady, Mayhem already weighed about fifty pounds and had an arm span of about twelve feet. Rocky was even larger. She was coy at first but then gave in to Rocky's advances—to the great interest of the assembled crowd at the aquarium and online viewers watching the live octolove webcam.

For most octopuses, this brief act marks the irrevocable beginning of their own demise. As much as sex is tied to the creation of new life, for the octopus it is also the start of death. For some species, the deed itself is marked with an ominous act of self-inflicted violence, with the male presenting his genetic load on a detached arm. So strange is this ritual that a nineteenth-century naturalist had seen these errant arms, which can still wriggle around, crawling on the surface of female argonauts. Thinking they were some kind of besuckered parasitic worm, he named it *Hectocotylus* for its hundred or so suckers. The name for the specialized member stuck.

These severed organs are also sometimes found *inside* females. And they were one way, before genetic analysis, that researchers could confirm that some super tiny males, such as those of the blanket octopus, are indeed the same species as the tremendous females that can outweigh them by so much.

The dark, campy film version of *The Spirit* even borrows from this capability, ending on a strangely sexual note. After an explosion is thought to have killed off Samuel L. Jackson's character, the Octopus, we see one severed finger (that is, arm) crawling toward his former assistant, Silken Floss (Scarlett Johansson). She spots the finger and keeps it, slipping it into her breast pocket, like a faithful octopus lover, plotting future generations.

Tentacle Erotica

The octopus's eight appendages have not escaped the notice of those with a libidinous view of the world. In fact, they have inspired an entire genre of erotica and even some of Picasso's work.

Scholars trace this "tentacle erotica" back at least to an early-nineteenth-century Japanese wood-block print, which is said to have been inspired by a legend of a female shell diver who must escape the clutches of a sea dragon god. In the process of her flight to freedom, she is pursued by a host of creatures, including octopuses. The Edo-period wood-block print is titled *Tako to Ama*, literally "Octopus and the Shell Diver" (commonly translated into English as "The Dream of the Fisherman's Wife," which puts a different spin on the whole thing altogether). It depicts a buxom woman reclined on her back among the rocks, with a large octopus performing cunnilingus and a smaller octopus kissing her mouth and wrapping an arm tip around a nipple. Their other arms wrap around her body, legs, and arms, and her hands grasp the girth of two of the large octopus's arms. The background is scribbled with text. Early Western scholars, who didn't translate the Japanese characters, took the scene to depict forced sex, but that interpretation is hard to support—even based on the image alone. And when you throw in some of the translations (woman: "your sucking at the mouth of my womb makes me gasp for breath! Aah! Yes, it's there! With the sucker, the sucker! Inside, squiggle, squiggle, ooh! Ooh, good!"), it obviously is a fantasy of mutual pleasure. Similarly scandalous scenes appear in other Japanese art and have even been carved into netsuke figurines.

Perhaps not quite as overtly sexual, a dynamic sculpture in Italy's Parco Montagnola, in Bologna, displays a suggestive octopus sculpture that startled me as I was wandering from the central train station to my nearby hotel. The octopus, in the bottom of the sculpture, is dwarfed by a

A detail of the grabby octopus in Bologna's famous sculpture.
(Katherine Harmon Courage)

powerful horse and busty virgin. But with its arms weaving through the virgin's thighs and reaching out for the horse, it appears strangely in control as it seems to drag them both down.

The exotic side of the octopus sneaked onto the covers of American pulp fiction and comic books in the mid-twentieth century. The self-proclaimed "Silly Website" Poulpe Pulps chronicles a sexy smattering of these works. It includes a cover from the 1930 book *The White Goddess* from the "Spicy Adventures" series, which features a smiling woman in a scanty-for-the-time red bathing suit with an octopus wrapping its arms suggestively around her. Another tantalizing illustration appeared on the 1940 manly *Argosy* title "Gateway to Oblivion," in which "Satan of the Sea Spreads Evil Tentacles to Guard the Treasures of the Deep." On its cover, a barely clad muscular young man wields a machete against an octopus that is beginning to enwrap him.

In Japan, modern-day tentacle erotica has continued in the artistic realm, notably by Toshio Saeki and Masami Teraoka. But it peaked in the folds of manga books and similarly styled animations, in which the act often turned violent and sometimes even involved robotic appendages. This harsher version is known as *shokushu goukan*, or "tentacle rape." Some scholars suggest that this offbeat fetish took off in Japan in part because of censorship rules that disallow depiction of the penis (but the octopus arm performing a similar task apparently was just fine). Given its biological hurdles, this genre remained fairly niched and didn't really take off in the live-action film world, although there are a few B-movie horrors in which fantastical creatures or other tentacle-like objects are used to perform sexual acts.

But, as with many things I did not expect to be doing when I first began this book, stumbling upon octopus porn on a Monday night after work was a bit of a surprise. But sure enough, it's out there. Just a collection of bare female pelvises with small, dead octopuses splayed on top, around, and yes, inside of them. Best not to try anything with a live octopus (in bed or the seabed), as its suckery grip is likely to leave you with dozens of telltale hickeys.

"Something Incalculable"

Back in the world of real octopus sex, male octopuses barely live long enough to brag of their conquests. After handing over their spermatophores—whether inserted into the female's funnel or handed over ceremoniously on a right leg—for reasons researchers still don't quite understand, the male octopus wanders off and enters a state of senescence and soon dies. Few of them live longer than a month after mating. In the wild, they will likely be eaten. And males kept in captivity slowly waste away, often losing some 20 percent of their body weight before dying.

After sex, the female stores the spermatophores in a nifty organ called the spermatheca until she is ready to lay her eggs. For females that might mate with more than one male, males compete to be the last one to mate with her before she spawns. As Janet Voight points out, "Frequently the last sperm in is the first one out" and thus has the better chance at fertilizing more of the eggs as they pass by. Scientists aren't sure how long females can keep the sperm, but one female giant Pacific octopus is reported to have laid fertilized eggs at an aquarium some seven months after being caught and kept sans male.

To lay their eggs, most females find a den that is suitably safe and secluded. Some female octopuses will even build walls of rock to blockade themselves inside with their brood. A female octopus can lay anywhere from 50 eggs to hundreds of thousands of them. Smaller eggs, which can be just a few millimeters long, are usually laid in larger batches. The common octopus can lay some 120,000 to 400,000 or more eggs that are just a little more than two millimeters long. The female tends these for about a month or two (depending on water temperature) until they hatch.

The octopus's oblong eggs are coated in chorion, an embryonic membrane, that is often pulled out to a thin chitin stem at one end. With this handy stem, females of most species can attach individual eggs either directly to the den wall or to one another in long strands that can be affixed to the wall in a unit. One rare video captured a female octopus using just the stalks of her suckers and her mouth to weave strands together to create long strings of eggs. Some female octopuses have been recorded working on this delicate project for a full month.

In Jean Painlevé's film *The Love Life of the Octopus*, rare footage of a female tending her egg strands is touching and elegant. The eggs appear like soft, feathery blossoms as the female fondles them continuously.

As Aristotle so poetically described the reproducing octopus: "Its spawn is shaped like a vine-tendril, and resembles the fruit of the white poplar; the creature is extraordinarily prolific, for the number of individuals that come from the spawn is something incalculable."

But this is just the beginning of the female octopus's maternal duties. From now on, the female will hardly eat. "The female octopus at times sits brooding over her eggs, and at other times squats in front of her hole, stretching out her tentacles on guard," Aristotle recounts. During this time, the female blows water over her eggs and actively grooms them to keep them clear of algae or other would-be colonizers. And because she is not eating much—if at all—during this period, the water flushed out of her funnel is clean and clear of waste.

Some species don't affix their eggs at all, but instead keep them together with their many arms. A female argonaut octopus lays her two-millimeter-long eggs in her thin, self-made shell. Blue-ringed octopuses often lay their eggs inside of other mollusks' shells—after the previous inhabitant has left, of course (though under what circumstances, it is not clear). Roy Caldwell, of the University of California, Berkeley, described one scary surprise after cracking open what he thought was an intact rock oyster shell. "My hammer cracked open an encrusted honeycomb oyster shell, and a tangle of arms covered with iridescent blue spots spewed from the fissure," he wrote in an article in *Freshwater and Marine Aquarium* magazine, recounting the encounter with the poisonous octopus. "Out charged a very irate golf ball–sized female octopus, holding in her arms a clutch of developing eggs. I remember being taken aback by her aggressive posture. Rather than crawling for cover like most octopus, she reared up while pulling back her first two pair of arms, exposing her mouth. It was very clear to me that here was an octopus ready to bite." It was most likely a *Hapalochlaena maculosa* or *Hapalochlaena fasciata*. For these species, "the brooding posture typically has the female sitting, mouth down, with her arms and web drawn up around her body, her eyes visible through the folds of the web," he wrote. "When forced to move, three pair of arms hold tightly to the eggs; one pair is used to crawl along the substrate."

Some octopus eggs, such as those of the aegina octopus (*Octopus aegina*), hatch after just a couple of weeks. Others, such as those of the giant Pacific octopus, which lays some seventy thousand eggs, can take six months to hatch. Scientists recently observed one outlier female *Graneledone boreopacifica* octopus in the cold depths of Monterey Canyon guarding a single brood of eggs for four and a half years.

In the weeks or months she spends tirelessly tending her eggs, the female can lose as much as 50 percent of her body mass—an effective, if drastic, postpregnancy diet.

But the female octopus is not doomed, necessarily, to die of starvation.

The peculiarity of her death runs deeper than that. Researchers have found that it is a genetically programmed suicide that actually kills these new-mother cephalopods. It starts in the optic glands, which release an endocrine that cues the cascade of biochemical changes that eventually lead to death. Many pet octopuses have had these glands removed so that they'll live longer. But without these glands, they lose ther maternal inclinations.

Even initially dismissive Aristotle seemed almost touched by this predestined demise. "After the eggs are laid, they say that both the male and the female grow so old and feeble that they are preyed upon by little fish, and with ease dragged from their holes," he wrote in the *History of Animals.* After mating, "the male becomes leathery and clammy," he noted. And "the female after parturition is peculiarly subject to this colliquefaction; it becomes stupid; if tossed about by waves, it submits impassively; a man, if he dived, could catch it with the hand; it gets covered over with slime, and makes no effort to catch its wonted prey."

The Next Generation

When the eggs are ready to hatch, as her final maternal act, sometimes the female will give them an extra push, blowing them out of the den and into the open ocean.

This event has a powerful effect on those who have witnessed it. Jean Boal has seen it more than once in her long research career, but she'll never forget the first time. Her lab at Millersville had one particularly large female that was at least fifteen months old—a few months past the typical year-long life span of her fellow lab octopuses.

Boal was working late into the evening in the lab, and the octopus was sitting with her eggs in a large overturned flowerpot. Usually the females would sit guarding the eggs, but this one had turned sucker-side out across the opening. She started "contracting and blowing and contracting and blowing," Boal recalls. "Every time she blew, all these babies came out into the water." With each pulse, hundreds of hatchlings were silhouetted against the light streaming in from street lamps through the darkened lab's windows, like little flecks of snow. It was so beautiful that she wanted to share it with someone, so she went running out into the hall, yelling for someone, anyone to come quick, to bring a camera. But no one was there. Everyone had left for the night, and Boal's calls echoed unanswered down the empty corridors. She returned to the tank, to watch the baby octopuses emerging. She had no photos of the occasion, "but I had my memory," she says.

These little octopuses are meant to be cast out into the wide world, so trying to keep them alive in a research lab or breeding program is difficult. Boal says she has been able to raise some hatchlings over the years. But these tiny animals present a surprisingly large space problem. The babies are okay swimming around in a large, well-appointed tank for a while, she says. But they grow so fast, and because they are occasionally inclined to eat one another, if you manage to keep forty-five octopuses alive, pretty soon you're going to need forty-five individual tanks.

In the wild, however, the new hatchlings are on their own as the mother octopus fades into obscurity. This parent-offspring separation makes so much of the octopus's complex behavior all the more impressive. So many species that we admire for their intellectual abilities—chimps, whales, elephants—spend an extended period of time being reared by their parents and others, who show them the ropes and teach them tricks. But each short-lived octopus must figure out these skills for itself. "They have to discover it from scratch," as Jennifer Mather says.

Hatchlings that come from small broods of large eggs—those that are closer to an inch or inch and a half long—often emerge looking almost like minioctopuses. These juveniles, including some species of superpoisonous blue-ringed octopuses, are ready to start life crawling along the seafloor as fully benthic beings.

More commonly, however, octopuses hatch from small eggs in batches of thousands of siblings and are cast out into the currents, where they float around as paralarvae—sometimes for months. They use their statocysts to stay oriented and their water jets to get around in the water column. But they are largely at the mercy of the currents.

Although these small cephalopods are unlikely to survive into adulthood, they scatter to farther reaches than those species that hatch from larger eggs and immediately start life on the seafloor. The drifting hatchlings might be vulnerable, but just a few survivors can spread their species and their genetic lines far afield.

A paralarva hatchling carries with it the remainder of an egg sac that might sustain it for a day or so, but then it needs to start finding food on its own. To do that, it uses its stubby arms to capture other larvae—such as those belonging to fish or shrimp—or small copepods. And feed, they do. A baby octopus can balloon in size about 5 percent each day.

Few will survive beyond this first perilous stage of life. These newly hatched octopus paralarvae are but tiny flecks of protein floating in the water column, mixing with other plankton and vulnerable to most any hungry animal larger than they are—including other octopuses, even their brood mates. Most will end up as food for fish, whales, jellyfish, and arrow worms. For deep-sea species, the challenge of staying alive also involves being able to locate enough food in the vast seas.

Common octopus larvae might bob around with fellow platonic friends—and foes—for a month or two, depending on the water temperature (with chillier waters leading to longer maturation periods). When the young octopus has grown large enough to live on the bottom (whether that has taken weeks or months), it changes from being attracted to light—which helps keep it swimming aloft in the water column—to being enticed by the darkness of safe hiding places on the floor. A common octopus might start to settle downward once it's about twelve millimeters long (less than half an inch). And it will find itself on the ocean floor and looking for love just a year or two after hatching, by which point it might already weigh more than four pounds.

In a clutch of tens of thousands of eggs, less than 0.003 percent might survive to reproduce and live through to their own genetically programmed death. But so long as roughly one male and one female make it through the mating process, an octopus population can sustain itself indefinitely.

And if a few make it into laboratories and aquariums and into the view of watchful researchers in the field, they might have the chance to teach us a little about what they know.

Epilogue

Over the millennia, our taste for octopus—scientifically, culturally, and culinarily—has continued to grow. And as we dive farther and deeper into the world's oceans and our encounters with these curious cephalopods proliferate, they will surely draw even more curiosity than they provide answers.

We should embrace this ongoing exploration. "A lot of basic research might not fulfill an immediate purpose for itself," Michael Kuba notes. "But it is a part of our culture to be curious and to want to acquire knowledge." As one of the most cognitively capable invertebrates, the octopus offers us a unique opportunity to study behaviors that have emerged as similar to those of vertebrates, yet from an evolutionarily alien path.

The study of the octopus, however, has actually become more difficult as scientists learn more about this variable invertebrate. As researchers are able to look, for the first time, into the DNA of these animals, the octopus family tree is becoming even more muddled. What were thought to be single unified species are turning out to be related subspecies. The giant Pacific octopus, for example, might actually be three distinct subspecies, one off the coast of Japan, one off the coast of Alaska, and another that lives near Puget Sound.

Roland Anderson, who has worked with this species for decades, says he's yet to be convinced. "I, myself, do not believe in subspecies, because I was taught that a species is immutable—unless it mutates into a new species," he says. "I am a lumper not a splitter." But he says if enough genetic evidence comes to light, he's willing to reconsider the existence of fissures in his beloved giant Pacific octopus species.

Even trusty old *Octopus vulgaris*, the common octopus, is looking like it might consist of more than one species now that we can survey its genetic profile across the globe. Nikolaos Schizas and his former graduate student Alexandre Jassoud studied the genetic structure of global common octopus populations. When they mapped the samples across the globe, common octopuses in the Mediterranean turned out to be more closely related to the Japanese specimens than they are to the Caribbean common octopuses. "We have a big difference between the Caribbean and the rest," he points out. "That's a big story," he says, because it hadn't yet been described formally, that is, genetically, with both mitochondrial and nuclear DNA. These differences are especially curious because even concerted morphological work had not been able to distinguish among different common octopuses from the world over. Janet Voight had looked two decades ago for differences in various common octopus specimens and not turned up much, says Schizas. "It's not unheard-of that morphology and molecules don't say exactly the same story," he says.

So our friend the common octopus might really be a complex of closely related subgroups rather than separate species. Perhaps we've arrived on the scene—with our fancy newfangled genetic sequencers—just in time to catch these octopuses in the midst of a speciation event, to witness it (in slow, slow motion) actually *becoming* two (or more) different species.

The speciation confusion also goes the other way. As Lydia Mäthger notes, "People still look at species that have been described in the lab, and they still give them new names because they think it's slightly different."

Genetics, however, are not yet able to paint the entire picture, at least not one that we can confidently follow. To truly categorize octopus species specifically we still need morphology—a physical description—too. "They are a very difficult group of animals to clinically describe," Eric Hochberg says, not in the least because they are so malleable, in their shapes and colors. So that means looking a little more closely than you might have to for a bird. "You have to use lots of elements of internal anatomy, the features of the skin, sucker counts," he says. Frequently, beautiful photos of mystery octopuses reach him and his colleagues, but without a physical specimen (or two or three of each sex) to measure and describe, they are at a loss. And unless the animal is fresh out of the water, the task of categorizing it becomes even more challenging. "Often the specimens have resided in collections and often have not been clinically fixed and preserved. So, here, unlike with squid, where you can easily see the body and the fins, you'll often get a scrunched-up—at best—octopus." Nevertheless, he does not discount these ready-and-waiting subjects. There are "whole populations of specimens and collections that need to be studied critically," he says.

A poorly preserved octopus, of course, beats no octopus at all. For the biggest octopuses, such as the giant Pacific octopus, Hochberg suggests that it's possible "there are several species that have all been lumped under the same name." That, in part, might be because they are so tough to get back to the lab. "Those animals are so large when they are mature, that trying to get the collection containers where you can put a two-hundred-pound octopus has been difficult." Even if a report comes in from the field that researchers are pretty sure they have discovered a different species from, say, the Aleutian Islands, if Hochberg and others have no specimens they can use to describe it, it's not much use to anyone.

On the other side of the size scale, he notes, "it's a problem with the little tiny pygmy octopuses that for years and years have been considered to be juveniles of other octopus species." More than a decade ago, Hochberg was working on a project to characterize octopuses on the west coast of North America. He was measuring and dissecting small octopuses when he realized he was not looking at just little young things after all. "I began to find out that an awful lot of them are fully mature, but they'd be maybe less than a gram—or less than an inch in total length," he says.

But recruiting new taxonomists to continue this sort of work lately has been challenging. Scientific fields from protozoa to plants to primates are suffering because fewer and fewer biology students choose to go into the field of categorization. And without these librarians of the natural world, we won't get very far in deciphering all that our planet has to offer. "There is just still an awful lot of work," says Hochberg, who retired from his post at the Santa Barbara Museum of Natural History in 2012. "You really need some younger people that are interested around the world in taxonomy—that are willing to do all the other basic morphology studies, then combine it with molecular studies."

Learning more about octopuses, of course, is not only to satisfy our own curiosity and desire to catalogue the world. They have a lot to lend us as well. "Mollusks are a pharmacological treasure trove," Frank Grasso says. "We haven't begun to scratch the surface of what we, as humans, could learn from all mollusks." He rattles off the list of other fields to which they could contribute: engineering, actuation, pharmacology, neurochemistry, brain design and control. "If that isn't enough for you, then something is wrong with you," he says, laughing.

Despite a surge in interest in these animals, research itself is ebbing. The biggest challenge of studying octopuses in the wild, Jennifer Mather jokes, is that we don't breathe water. "If we want to know what they're doing, we have to go to their environment." And diving to search for these elusive animals is time-consuming and labor intensive, and thus quite expensive.

For that reason, we continue to miss out on understanding most of the world. We tend to study what is convenient (backyard birds!) and affordable to fund (field ecology!). But to grasp the world that so many of the planet's organisms live in—the oceans—we must travel there. "All of our best ideas have come from field work where you can't control" everything, Roger Hanlon says of his lab's research. "That's nature—go see how it works."

Destinations Unknown

As a short-lived, widespread, and, apparently, adaptable animal, the octopus doesn't face nearly the same level of threat from modern humans, with our furious-fishing and climate-changing habits, as does other ocean life, such as whales, tuna, or coral.

But it is not immune to our encroachments into the oceans. To be sure, the octopus has profited from some of our underwater development—and trash—making homes in pier pilings, sunken ships, and discarded beer bottles. "Oh, our litter has been wonderful," says Roland Anderson, who has seen countless octopuses living in and around our refuse in his decades of diving. And Joshua Rosenthal reveals that when he worked in Monterey and they needed to collect octopuses, instead of going diving for them, they would tie a bunch of empty wine or beer bottles to a long string and sink it off the coast for a day or two. Sometimes every single bottle would come up with an octopus in it.

Some research suggests that human fishing pressure on large finfish has actually helped octopus populations by removing predators and competitors. But within the sciences there is disagreement on how to look at these relationships. As Jaime Otero explained as we were driving along the Galician coast, there is a rift between many fisheries biologists and marine ecologists. Fisheries biologists are often focused on surveys of specific species. And marine ecologists are prone to puzzling over the broader questions, such as what happens to copepod populations when cod are depleted and how will *that* affect the overall food web, including the octopus populations? And then what might happen to *other* populations with a cephalopod population shift? Our fisheries models, no matter how detailed, have only just scratched the surface of these complex connections.

Rapid climate change, too, will no doubt have some effect on octopuses, their environments, and their food supply. Otero notes that even in the prolific, *pulpo*-centric region of Galicia, the intensity of the crucial upwelling effect has been dropping off (by some 45 percent) in the past decades. And in the well-studied Hood Canal in Puget Sound, which is home to many giant Pacific octopuses, annual dead zones have grown larger in recent years.

Although some scientists have predicted that octopuses might flourish in warmer waters, a recent analysis says otherwise. Jessica André, of the University of Tasmania, modeled the effect of climate change on Western Bass Strait pale octopus (*Octopus pallidus*) populations through the year 2070. Her data suggest that these changes will cause life spans to shrink and make groups less likely to survive large-scale catastrophes.

Their sensitivity to changing pH levels in the oceans is also "a huge concern," Roger Hanlon says, especially in the face of the ocean acidification that comes as a result of extra carbon dioxide in the atmosphere. Because octopuses rely on hemocyanin rather than hemoglobin in their blue, copper-based blood, a small shift in pH can reduce their ability to carry oxygen throughout their bodies.

"I learned this the hard way," Hanlon says. "An octopus can take all kinds of what we'd call dirty water. But they cannot take water in which the pH drops. A fish can take a significant pH drop from 8.0 all the way down to 5.5—and that's an exponential scale. A cephalopod, if it just goes down a few tenths of a point, from 8.0 to 7.7 or 7.5, that animal is in serious jeopardy—life and death." Perhaps the sensitive cephalopods will serve as indicator species, the canary in the acidifying ocean.

Acidification has already been shown to be stunting small-animal shell growth by hampering the calcium carbonate formulation. Because an octopus's livelihood depends on eating so many shelled species, the cephalopod hunter might suffer down the road—even if it gets a few easier meals at the outset.

Octopuses are also sensitive to the changing tides of weather events. During El Niño, for example, the octopuses along the southern California and Mexican coastline come in closer to shore, leaving them even more accessible to fishing.

And, of course, as octopus populations change, so, too, will those of their predators. Because the endangered Hawaiian monk seal, for example, feeds on local octopus populations, some conservationists have recommended keeping a close eye on octopus populations so as not to deprive the already struggling seal population of important food.

Our own dip into the octopus population for food is not insignificant. At first I thought it was just me—and the restaurants I often found myself seated in. But a couple of years ago, I started seeing octopus everywhere: a smoked octopus à la barca on my old Manhattan corner at The Dutch; a mashed potato and *pulpo* dish at Ñ, on Crosby Street. But it was as I was sitting down to a dinner at Café Select, a cozy Swiss-modern bistro on Lafayette Street, where my husband asked if I had read the menu's appetizer list—which featured octopus salad with celery, cherry tomatoes, and fennel—that I realized that the octopus fetish was beginning to go too far. Octopus was now climbing into the cuisine of the Alps by way of SoHo. Soon it was even popping up in front of me on pizzeria menus from Bridgehampton to Bed-Stuy. And that was just in New York. In my travels, I encountered octopus spaghetti in Malta, octopus ceviche in Puerto Rico, and even octopus soppressata in Italy. Can this octomania possibly be sustainable?

In many places, such as Greece, where octopus fishing—and eating—has been a central part of the culture for centuries, if not millennia, local populations have taken a big hit. Takis Zaloumis in

Octopus soppressata in Italy. (Katherine Harmon Courage)

Gythio notes that he has seen the amount of octopus caught locally decline "little by little" over the years. And it's not for want of effort. In fact, he blames the *increase* in the number of fishermen.

With serious fishing pressures—which have meant fewer and smaller octopuses—interest in aquaculture has increased. Attempts to farm octopuses in captivity, however, have been less than a rousing success so far. Getting them from the paralarvae stage to that of culturable juveniles has been tricky, as the youngest octopuses insist on eating live meals and aren't going to take frozen bits of fish for food. Efforts, from Spain to Italy to China to Japan, are underway to try to figure out how to culture these tough-to-grow animals.

We still have next to no idea just how much of an impact our growing appetite for octopus is having on their populations in the wild. If big fish stocks, such as cod or tuna, are tricky to track, reclusive octopuses are nearly impossible. To measure populations of fish, researchers can use hydroacoustics, which employs sound waves to find and measure masses of animals underwater. This method is an efficient way to get a quick count of a school of fish. But even squid, which live in the open water, are tricky to track this way because they are often found among schools of tasty fish and have a sound signature very similar to the fish. But with seafloor life like octopuses, which prefer to crawl alone among the rocks at the bottom, there is just no measuring them this way.

For schooling fish, governments and other regulatory bodies also know approximately how many nets are cast and how many pounds are hauled in. Fisheries scientists can catch individual fish and check their age by counting the rings on an ear bone known as an otolith, much as one can count the rings on a tree trunk. This demographic information can help paint a picture of the population. Based on years of this sort of data, scientists can make what they call stock assessment models of how fish populations are expected to do. But this is no easy task, even with well-tracked species. While working in fisheries research, Otero once had an office mate who had previously been an astrophysicist. "He was fed up with solar physics," so he decided to try his hand at fisheries modeling, Jaime Otero says. The astrophysicist was stunned by the difficulty, finding fisheries mathematics to be, as he had put it wryly, "complex." To say the least. And "models for finfishes don't apply to octopuses," Otero adds.

Ángel Guerra and others have been eyeing the hard, bonelike structure that helps octopuses orient themselves in the water (the statolith) as a clue to age. Like the fish's otolith, the statolith grows over time, depositing layers as it grows. But in octopuses and cuttlefish, the layers seem to be gaining new layers irregularly. Now, however, researchers have been looking to another internal structure known as the stilet (a remnant of the octopus's ancient ancestor's shell). Until recently, scientists weren't sure how quickly it expanded and how frequently new layers were deposited. The rings, though, are so faint that they are incredibly difficult to count. But Guerra and his co-workers showed that, at least in adult octopuses, these layers seem to be deposited daily. So, he says, that would mean that even an octopus that weighed eight kilograms might not be much older than two years by ring count. "They grow very quickly—very quickly," he says. So the biggest one plucked from that region of Spain—by an unsuspecting fisherman with just a fishing line—at about thirty-five pounds, might not have been more than a few years old.

With many traditional tracking or counting methods off the table, scientists have relied on catch numbers to estimate the health of stocks. The animals are so elusive, though, that some areas, such as Alaska, have considered just getting a bead on the local habitat and using that to estimate how many giant octopus *might* be hiding beneath the dark waves.

A global dearth of population—or even catch—information is a major stumbling block for people like Otero, who could help to create management plans to keep long-productive populations like those around Vigo alive and thriving. And in places with even less information and active management, such as Mauritania and Vietnam, the situation is that much more precarious.

Of course, the octopus has survived for these many millions of years, through major extinction events and climate changes. So it, along with other smart cephalopods, will quite possibly outlive us on this changing planet. Studying its evolutionary path here might help us understand how it—and the rest of the oceans' creatures—will fare in coming years.

But getting to know the octopus's future might be almost as problematic as figuring out its past. As Joanne Kluessendorf points out, not only is "the fossil record very incomplete when it comes to soft-bodied organisms," but "it's almost impossible to expect to find a complete geologic record of them," because they fossilize so infrequently. Even the hard beaks, which are made of chitin, do not take to fossilization as easily as bone. Thus for now we are left with the 296-million-year-old *Pohlsepia* and the 95-million-year-old specimens from Lebanon to puzzle over.

Genetic and phylogenetic analyses of living octopus species are helping us to get a picture of some of the animal's evolutionary past. "We do really well with DNA when it comes to more recent ancestors, but it stops dead when you're really trying to look way back in time," says Kluessendorf. So to go back further, we need more fossils. "That's not to say that sitting in a collection in somebody's home or in a museum drawer there isn't an octopus" fossil waiting to be discovered and described, she says. If she were going to hunt for octopus fossils, she says, rather than hitting the field, "I would look in drawers that were labeled *'problematica'* or *'enigmatica,'*" dumping grounds for odd objects that earlier naturalists did not yet know how to categorize. "Indeed, there could easily be fossilized octopuses" there waiting to be discovered tomorrow, she says. After all, there are so many specimens out there—fossilized or more recently preserved by collectors—that haven't been looked at in decades or even centuries. "There's only so much a generation of scientists can do," she says, "and it just may not be the generation of the octopus right now."

The key, though, she says, is to be looking—to always be looking. Fossilized octopuses might be even more difficult to spot than a live one that's camouflaging in the wild. Even the best fossil specimens are often little more than a "colored stain on a rock," Kluessendorf says. "It's hard to train your eye to be looking for them."

But look we must—for the ancient ones, the deep-sea ones, and the tiny hatchling ones. Look we must for the mysterious and the familiar. For even if that strange brain, those eight amazing arms, those smart and dexterous suckers, those short generations, and that seemingly insatiable curiosity always remain just beyond the limits of our understanding, at least we will continue to discover and appreciate that which Jean Painlevé calls the joyous confusion, the unknown and the miraculous.

Illustration of *Octopus verrucosus* from 1886 by William Evan Hoyle.

ACKNOWLEDGMENTS

This book never would have come into existence if it were not for a lingering, long-unopened note from a mysterious stranger. Thank goodness I finally read it, because that message was the genesis of this book, and that unknown person turned into my fabulously amazing agent, Meg Thompson, of the Einstein Thompson Agency. Thank you, Meg, for finding me, for finding this wonderful idea, and for finding my outstanding publisher, Current. Thank you to my first two editors there—Courtney Young, who took a chance on octopuses, and Jillian Gray, who gave patient first edits—and to my final editor, Maria Gagliano, for taking the octopus by all eight arms and wrestling it to the finish line. Thank you as well to Eric Meyers, Bria Sandford, and Julia Batavia for keeping the project on track throughout all of the changes.

Unending gratitude to my mom, Pam, and my dad, Bill, for laying the necessary groundwork of boundless curiosity, a strong work ethic, and just enough whimsical craziness necessary for a book project (especially one on a topic one knows next to nothing about at the outset). A big thanks to my brothers, Andrew and Adam, for being so darn cool and helpful. Thank you, as well, to my grandmother Adele Rogers, for sharing your continuous love and fortitude with me. I could never have done any of this without the lifelong support and inspiration from my grandfather Ted Rogers. Ted, thank you for sending all of that wacky Russian literature to me in Tulsa when I was a kid. And for absolutely everything else. And thank you to Betsy Rogers, another font of encouragement and brilliance in my life.

To David Courage, my running partner, my first line editor, and my husband, this book—and my life—would be far worse off without you. Somehow you met me, moved in with me, and still decided to marry me all while I was (okay, *we* were) living under the reign of the octopuses. Thank you for sticking by a girl who always needs to do "just a little more octopussing."

And to my patient, patient friends, thank you for feigning interest in "octopi" for so many months over the course of so many dinners, drinks, and brunches. I'll try to pick something less awkward to pluralize next time around.

Thank you to my colleagues at *Scientific American* for teaching an English major so much about science and about journalism. I am especially grateful for the support and encouragement from my editors: Robin Lloyd, Phil Yam, and Mariette DiChristina. Thank you for letting me do the ridiculous blog post about tool-wielding octopuses that got this whole mess started.

Gratitude to all of my amazing writing instructors through the years: Mark Johnson, Derick Perry, Peter Robinson, Mark Amodio, Frank Bergon, John Schneller, and all of the other inspiring teachers at Holland Hall, Thacher, Vassar, and the Missouri School of Journalism.

Emphatically, I am indebted to each and every person I spoke with for this book. The breadth of your knowledge and the generosity of your time truly bowled me over. Thank you, Roland Anderson, Roy Armstrong, Richard Baraniuk, Jean Boal, Marcello Calisti, Matteo Cianchetti, Vinny Cutrone, Tamsen DeWitt, José Dios, Sandra Garrett, Francesco Giorgio-Serchi, Ángel Gonzalez, Frank Grasso, Ángel Guerra, Roger Hanlon, Naomi Halas, Eric Hochberg, Binyamin Hochner, Joanne Kluessendorf, Michael Kuba, Cecilia Lasch, Stephan Link, Manuel Manolo, Laura Margheri, Jennifer Mather, Lydia Mäthger, Bruno Milone, Maria Fernanda Montiel, Tony Morelli, Jaime Otero, Joshua Rosenthal, Nikolaos Schizas, Janet Voight, James Wood, Takis Zaloumis, and Odile—and everyone else I met along the way. A big Greek *efharisto* to our ancient octopus admirer, Aristotle. And *merci* to the works, words, and wisdom of Jean Painlevé.

And to the octopuses, thank *you*! My deepest apologies to those who were eaten for the sake of research. I wish you weren't so nutritious and delicious, but you are. So, reading public, thank you for your support—and for not letting their sacrifice have been in vain!

NOTES

Introduction

3 **We currently drag in:** Stelio Katsanevakis and George Verriopoulos, "Seasonal Population Dynamics of *Octopus vulgaris* in the Eastern Mediterranean," *International Council for the Exploration of the Sea*, 2005. http://icesjms.oxfordjournals.org/content/63/1/151.full.pdf.

5 **It is "chunky in appearance":** "Species Fact Sheets, *Octopus vulgaris*," Fisheries and Aquaculture Department, Foods and Agriculture Organization of the United Nations. http://www.fao.org/fishery/species/3571/en.

5 **mostly along continental shelves:** Ibid.

6 **earliest recognizable forebears:** "Cephalopoda," University of California Museum of Paleontology. http://www.ucmp.berkeley.edu/taxa/inverts/mollusca/cephalopoda.php.

6 **the Carboniferous seas:** Joanne Kluessendorf and Peter Doyle, "*Pohlsepia mazonensis*, an Early 'Octopus' from the Carboniferous of Illinois, USA," *Palaeontology*, 2003. http://onlinelibrary.wiley.com/doi/10.1111/1475-4983.00155/abstract.

6 **It's estimated to have lived:** Ibid.

6 **95-million-year-old fossilized:** Dirk Fuchs, Giacomo Bracchi, and Robert Weis, "New Octopods (Cephalopoda: Coleoidea) from the Late Cretaceous (Upper Cenomanian) of Hâkel and Hâadjoula, Lebanon," *Palaeontology*, 2008. http://onlinelibrary.wiley.com/doi/10.1111/j.1475-4983.2008.00828.x/abstract.

6 **"the octopus is a stupid creature":** Aristotle, *History of Animals*, Kindle edition, 273.

Chapter One: To Catch an Octopus

11 **people have been catching:** Jennifer Mather, Roland Anderson, and James Wood, *Octopus: The Ocean's Intelligent Invertebrate* (Timber Press, 2010), 67.

11 **a new Japanese contraption:** Gilberto Carreira and João Gonçalves, "Catching *Octopus vulgaris* with Traps in the Azores: First Trials Employing Japanese Baited Pots in the Atlantic," *Marine Biodiversity Records*, 2009. http://journals.cambridge.org/action/displayAbstract?fromPage=online&aid=5776520.

12 **The lures, now often made:** "Hawaiian Octopus *Octopus cyanea*," Seafood Watch Seafood Report, Monterey Bay Aquarium, 2004. http://www.montereybayaquarium.org/cr/cr_seafoodwatch/content/media/MBA_SeafoodWatch_HIOctopusReport.pdf.

12 **"The cuttle-fish, the octopus":** Aristotle, *History of Animals*, Kindle edition, 107.

12 **Some fishers in Mauritius:** "Octopus (including *Octopus cyanea*) Republic of the Philippines," Seafood Watch Seafood Report, Monterey Bay Aquarium, 2011. http://www.montereybayaquarium.org/cr/cr_seafoodwatch/content/media/MBA_SeafoodWatch_PhilippineOctopusReport.pdf.

12 **In Mauritian octopus trawls:** "Tako (Madako) Common Octopus, *Octopus vulgaris*," Seafood Watch Seafood Report, Monterey Bay Aquarium, 2008. http://www.montereybayaquarium.org/cr/cr_seafoodwatch/content/media/MBA_SeafoodWatch_TakoOctopusReport.pdf.

13 **Vietnamese fishermen keep:** Ibid.

13 **offshore muddy or sandy areas:** Ibid.

19 **starting a downwelling cycle:** X. A. Alvarez-Salgado et al., "Surface Waters of the NW Iberian Margin: Upwelling on the Shelf Versus Outwelling of Upwelled Waters from the Rias Baxias," *Estuarinen, Costal and Shelf Science*, December 2000. http://www.sciencedirect.com/science/article/pii/S0272771400907145.

30 **has played out right in front:** Frank Lane, *Kingdom of the Octopus: The Life History of the Cephalopoda* (Sheridan House, 1960), 50.

30 **A video from the Seattle:** "Mystery: Sharks missing at Seattle Aquarium." http://www.youtube.com/watch?v=urkC8pLMbh4.

31 **Atlantic violet blanket octopus:** Everet Jones, "*Tremoctopus violaceus* Uses *Physalia* Tentacles as Weapons," *Science*, 1963.

31 **University of Arizona marine:** Rafe Sagarin, *Learning from the Octopus: How Secrets from Nature Can Help Us Fight Terrorist Attacks, Natural Disasters, and Disease* (Basic Books, 2012).

36 **Back in 1963:** Malcolm R. Clarke, "Economic Importance of North Atlantic Squids," *New Scientist*, March 1963.

36 **Galician fishermen alone:** Francisco Rocha, Ángel Guerra, Ricard Prego, Uwe Piatkowski, "Cephalopod Paralarvae and Upwelling Conditions Off Galician Waters (NW Spain)," *Journal of Plankton Research*, 1999. http://plankt.oxfordjournals.org/content/21/1/21.full.pdf.

37 **Spanish were exporting:** Alberete Chapela, Ángel González, Earl Dawe, Francisco Rocha, Ángel Guerra, "Growth of Common Octopus (*Octopus vulgaris*) in Cages Suspended from Rafts," *Scientia Marina*, 2006. http://www.abdn.ac.uk/CIAC/ongrowing.pdf.

37 **In the late twentieth:** Seafood Watch Seafood Report, 2008.

37 **launched a long-term:** Ibid.

37 **Mexico's Gulf of California:** "Gulf of California Seafood Report: Southwest Region," Seafood Watch Seafood Report, Monterey Bay Aquarium, 2007. http://www.montereybayaquarium.org/cr/cr_seafoodwatch/content/media/MBA_SeafoodWatch_GulfofCalifornia_Guide.pdf.

37 **Teams of two or three:** Ibid.

37 **shallower areas off the coast:** Ibid.
37 **traps in the rocky areas:** Ibid.
37 **don't bring in much more:** Ibid.
38 **much of its octopus from the Philippines:** "Octopus (including *Octopus cyanea*), Republic of the Philippines," Seafood Watch Seafood Report, Monterey Bay Aquarium, 2011. http://www.montereybayaquarium.org/cr/cr_seafoodwatch/content/media/MBA_SeafoodWatch_PhilippineOctopusReport.pdf.
38 **More than a dozen shallow-water:** Ibid.
38 **Most of the sushi-grade:** "Octopus, Common/Sushi (Worldwide, Wild-caught)," Seafood Watch, Monterey Bay Aquarium. http://www.montereybayaquarium.org/mobile/sfw/FishDetails.aspx?fid=259®ion_id=8.
38 **This processing stopover:** Seafood Watch Seafood Report, 2008.
38 **Vietnam, for example:** Ibid.
38 **Seafood Watch suggests that consumers:** Ibid.
39 **fished by local Moroccans:** Ibid.
39 **by the turn of the century:** Ibid.
39 **annual overall fishing quota:** Ibid.
39 **Mauritania also signed:** Ibid.
40 **Many saw the 25 percent:** Ibid.
40 **Seafood Watch suggests that:** Ibid.
40 **has already instituted size:** Ibid.
40 **octopus catch exploded:** Ibid.
40 **Since 2005, the European Union:** Ibid.
40 **Galician rules from at least:** Alberete Chapela, *Scientia Marina*, 2006.

Chapter Two: Tough and Tasty

47 **treatment so charmingly espoused:** Irma Rombauer and Marion Rombauer Becker, *Joy of Cooking: The All-purpose Cookbook* (Plume, 1997), 408–9.
55 **In a 1935 essay:** Jean Painlevé, "Feet in the Water," *Voilà*, 1935. Republished in *Science Is Fiction: The Films of Jean Painlevé*, edited by Andy Masaki Bellows and Marina McDougall (Brico Press, 2000), 136–38.

Chapter Three: A Strange Animal

63 **This genus captivated:** Aristotle, *History of Animals*, Kindle edition, 274.
65 **These gals can grow:** Mark Norman, David Paul, Julian Finn, and Tom Tregenza, "First Encounter with a Live Male Blanket Octopus: The World's Most Sexually Size-Dimorphic Large Animal," *New Zealand Journal of Marine and Freshwater Research*, 2002. http://www.selfishgene.org/Tom/Papers/MDNetal_NZJFMR02.pdf.
66 **in one seaside studio:** Jean Painlevé, "Feet in the Water," *Voilà*, 1935. Republished in *Science Is Fiction: The Films of Jean Painlevé*, edited by Andy Masaki Bellows and Marina McDougall (Brico Press, 2000), 132.
67 **Even back in the fourth century:** Aristotle, 89.

68 **when offered an object:** Ruth Byrne, Michael Kuba, Daniela Meisel, Ulrike Griebel, Jennifer Mather, "Does *Octopus vulgaris* Have Preferred Arms?" *Journal of Comparative Psychology*, 2006. http://www.ncbi.nlm.nih.gov/pubmed/16893257.

69 **Zoologist Martin Wells:** Martin Wells, "The Heartbeat of *Octopus vulgaris*," *Journal of Experimental Biology*, 1979. http://jeb.biologists.org/content/78/1/87.full.pdf.

71 **He noted in a 1983 paper:** Eric Hochberg, "The Parasites of Cephalopods: A review," *Memoirs of the National Museum of Victoria*, 1983.

71 **To that last location:** Eric Hochberg, "The 'Kidneys' of Cephalopods: A unique habitat for parasites," *Malacologia*, 1982.

73 **reliably recorded:** James Cosgrove, "Aspects of the Natural History of *Octopus dofleini*, the Giant Pacific Octopus," University of Victoria, 1987.

73 **One dead massive:** Steve O'Shea, "The Giant Octopus *Haliphron atlanticus* (Mollusca: Octopoda) in New Zealand Waters," *New Zealand Journal of Zoology*, 2004.

74 **in the Inland Sea of Japan:** "Species Fact Sheets, *Octopus vulgaris*," Fisheries and Aquaculture Department, Foods and Agriculture Organization of the United Nations. http://www.fao.org/fishery/species/3571/en.

74 **Lab experiments showed:** "Hawaiian Octopus *Octopus cyanea*," Seafood Watch Seafood Report, Monterey Bay Aquarium, 2004. http://www.montereybayaquarium.org/cr/cr_seafoodwatch/content/media/MBA_SeafoodWatch_HIOctopusReport.pdf.

74 **For Voight's dissertation:** Janet Voight, "Population Biology of *Octopus digueti* and the Morphology of American Tropical Octopods," University of Arizona, 1990. http://arizona.openrepository.com/arizona/handle/10150/185018?mode=full.

74 **A common octopus fella:** "Tako (Madako) Common Octopus, *Octopus vulgaris*," Seafood Watch Seafood Report, Monterey Bay Aquarium, 2008. http://www.montereybayaquarium.org/cr/cr_seafoodwatch/content/media/MBA_SeafoodWatch_TakoOctopusReport.pdf.

74 **The gals get a little:** Ibid.

76 **It has since become, as biologist:** Jennifer Mather, Roland Anderson, and James Wood, *Octopus: The Ocean's Intelligent Invertebrate* (Timber Press, 2010), 73.

78 **In cold temperatures:** Sandra Garrett and Joshua Rosenthal, "RNA Editing Underlies Temperature Adaptation in K+ Channels from Polar Octopuses," *Science*, 2012. http://www.sciencemag.org/content/335/6070/848.

Chapter Four: Skin Tricks

86 **tend to have big eyes:** Jennifer Mather, Roland Anderson, James Wood, *Octopus: The Ocean's Intelligent Invertebrate* (Timber Press, 2010), 105.

86 **takes advantage of this:** Ibid.

87 **species will hover near:** Ibid., 93–94.

88 **ocean explorer David Gallo:** "David Gallo: Underwater Astonishments," TED, 2007. http://www.ted.com/talks/david_gallo_shows_underwater_astonishments.html.

91 **octopuses are, in fact, picking:** Noam Josef, Piero Amodio, Graziano Fiorito, and Nadav Shashar, "Camouflaging in a Complex Environment—Octopuses Use Specific Features of Their Surroundings for Background Matching," *PLOS ONE*, 2012. http://www.plosone.org/article/info:doi/10.1371/journal.pone.0037579.

92 **the South American catfish:** Markos Alexandrou, Claudio Oliveira, Marjorie Maillard, Rona McGill, Jason Newton, Simon Creer, and Martin Taylor, "Competition and Phylogeny Determine Community Structure in Mullerian Co-mimics," *Nature*, 2011. http://www.nature.com/nature/journal/v469/n7328/full/nature09660.html.

92 **The mimic octopus:** Eric Hochberg, Mark Norman, and Julian Finn, "*Wunderpus photogenicus* n. gen. and sp., A New Octopus from the Shallow Waters of the Indo-Malayan Archipelago (Cephalopoda: Octopodidae)," *Molluscan Research*, 2005. http://www.mapress.com/mr/content/v26/2006f/n3p140.pdf.

92 **In 2011, Hanlon:** Chuan-Chin Chiao, Kenneth Wickiser, Justine Allen, Brock Genter, and Roger Hanlon, "Hyperspectral Imaging of Cuttlefish Camouflage Indicates Godo Color Match in the Eyes of Fish Predators," *Proceedings of the National Academy of Sciences*, 2011. http://www.pnas.org/content/early/2011/05/10/1019090108.full.pdf.

95 **Iridophores also reflect:** Lydia Mäthger and Roger Hanlon, "Anatomical Basis for Camouflage Polarized Light Communication in Squid," *Biology Letters*, 2006. http://www.ncbi.nlm.nih.gov/pmc/articles/PMC1834008/.

96 **one species of deep-sea:** Sonke Johnsen, Elizabeth Balser, Eric Fisher, and Edith Widder, "Bioluminescence in the Deep-Sea Cirrate Octopod *Stauroteuthis syrtensis* Verrill (Mollusca: Cephalopoda)," *The Biological Bulletin*, 1999. http://www.biolbull.org/content/197/1/26.abstract.

98 **A little spurt of the neurotransmitter:** John Messenger, C. Cornwell, and C. Reed, "L-Glutamate and Serotonin Are Endogenous in Squid Chromatophore Nerves," *The Journal of Experimental Biology*, 1997. http://jeb.biologists.org/content/200/23/3043.full.pdf.

100 **A 2011 study showed:** Daisuke Kojima, Suguru Mori, Masaki Torii, Akimori Wada, Rika Morishita, and Yoshitaka Fukada, "UV-Sensitive Photoreceptor Protein OPN5 in Humans and Mice," *PLOS ONE*, 2011. http://www.plosone.org/article/info%3Adoi%2F10.1371%2Fjournal.pone.0026388.

107 **one of these multiyear:** "Scientists Dive into Study of Squid Skin Sensors," *Biophotonics*, 2011. http://www.photonics.com/Article.aspx?AID=47723.

109 **jump in the detection direction:** Mark Knight, Heidar Sobhani, Peter Nordlander, and Naomi Halas, "Photodetection with Active Optical Antennas," *Science*, 2011. http://www.sciencemag.org/content/332/6030/702.abstract.

Chapter Five: Brain Power

113 **an experiment for a female:** Roland Anderson, "Smart Octopus?" *The Festivus*, 2006. http://www.thecephalopodpage.org/OctopusSmarts.php.

114 **"the trouble lies in actually":** Anderson.

120 **Aristotle's famous quip:** Aristotle, *History of Animals*, Kindle edition, 273.

122 **giant Pacific octopus, described by:** Jennifer Mather and Roland Anderson, "Exploration, Play and Habituation in Octopuses (*Octopus dofleini*)," *Journal of Comparative Psychology*, 1999. http://psycnet.apa.org/index.cfm?fa=buy.optionToBuy&id=1999-03911-013.

125 **Boal's undergraduate students:** Marie Beigel and Jean Boal, "The Effect of Habitat Enrichment on the Mudflat Octopus," *The Shape of Enrichment*, 2006. http://www.millersville.edu/biology/faculty/boal-pdf/4.beigel_boal_enrichment_2006.pdf.

127 **inspiration for the Beatles':** John Lennon, Paul McCartney, George Harrison, and Ringo Star, *The Beatles Anthology* (Chronicle Books, 2000).

127 **One high-tech:** "University of Colorado Archaeologist, Colleagues Hot on the Trail of Ancient Persian Warships," *EurekAlert!* 2004. http://www.eurekalert.org/pub_releases/2004-02/uoca-uoc020304.php.

128 **Some veined octopuses:** Julian Finn, Tom Tregenza, and Mark Norman, "Defensive Tool Use in a Coconut-carrying Octopus," *Current Biology*, 2009. http://www.cell.com/current-biology/abstract/S0960-9822(09)01914-9.

129 **octopus behavior in the 1980s:** Jennifer Mather, Roland Anderson, and James Wood, *Octopus: The Ocean's Intelligent Invertebrate* (Timber Press, 2010), 125.

130 **he realized that keepers:** Roland Anderson, "Cephalopods at the Seattle Aquarium," *International Zoology Yearbook*, 1987. http://onlinelibrary.wiley.com/doi/10.1111/j.1748-1090.1987.tb03132.x/abstract.

131 **one early personality test:** Jennifer Mather and Roland Anderson, "Personalities of Octopuses (*Octopus rubescens*)," *Journal of Comparative Psychology*, 1993. http://homepage.psy.utexas.edu/homePage/Group/AnimPersInst/Animal%20Personality%20PDFs/M/Mather%20&%20Anderson(1)%201993.pdf.

131 **Another study sought:** David Sinn, N. Perrin, Jennifer Mather, and Roland Anderson, "Early Temperamental Traits in an Octopus (*Octopus bimaculoides*)," *Journal of Comparative Psycholgy*, 2001. http://www.ncbi.nlm.nih.gov/pubmed/11824898.

133 **After just one year at:** Jeanette Scarsdale, "Octopus Home Again After Science Center Stint," *Kitsap Sun*, July 8, 2010. http://www.kitsapsun.com/news/2010/jul/08/octopus-home-again-after-science-center-stint/#axzz2Ht7ZvNyq.

133 **reported that octopuses that:** Graziano Fiorito and Pietro Scotto, "Observational Learning in *Octopus vulgaris*," *Science*, 1992. http://marinediscovery.arizona.edu/2005/2005%20Readings/Fiorito_octopus.pdf.

136 **A couple of Hawaiian:** John Forsythe and Roger Hanlon, "Foraging and Associated Behavior by *Octopus cyanea* Gray, 1849 on a Coral Atoll, French Polynesia," *Journal of Experimental Marine Biology and Ecology*, 1997. http://www.sciencedirect.com/science/article/pii/S0022098196000573.

137 **To test the octopuses:** Jean Boal, Andrew Dunham, Kevin Williams, and Roger Hanlon, "Experimental Evidence for Spatial Learning in Octopuses (*Octopus bimaculoides*)," *Journal of Comparative Psychology*, 2000. http://psycnet.apa.org/index.cfm?fa=buy.optionToBuy&id=2000-05103-004.

138 **Two thirds of the octopuses:** Boal et al.

142 **who coauthored the book:** Roger Hanlon and John Messenger, *Cephalopod Behaviour* (Cambridge University Press, 1996).

143 **A 1971 paper by Packard:** Andrew Packard, "Cephalopods and Fish: The limits of convergence," *Biological Reviews*, 1972. http://onlinelibrary.wiley.com/doi/10.1111/j.1469-185X.1972.tb00975.x/abstract.

143 **That's what Grasso:** Frank Grasso and Jennifer Basil, "The Evolution of Flexible Behavioral Repertoires in Cephalopod Molluscs," *Brain, Behavior and Evolution*, 2009. http://content.karger.com/ProdukteDB/produkte.asp?Doi=258669.

145 **As philosopher Thomas Nagel:** Thomas Nagel, "What Is It Like to Be a Bat?" *The Philosophical Review*, 1974. http://organizations.utep.edu/portals/1475/nagel_bat.pdf.

146 **in the UK, octopuses:** "Animals (Scientific Procedures) Act 1986," Chapter 14. http://www.homeoffice.gov.uk/publications/science-research-statistics/animals/transposition_of_eudirective/consolidated_aspa?view=Binary.

146 **The European Union has:** Nicola Nosengo, "European Directive Gets Its Tentacles into Octopus Research," *Nature*, 2011. http://www.nature.com/news/2011/110412/full/news.2011.229.html.

146 **And the 2012 Cambridge:** Philip Low, "The Cambridge Declaration on Consciousness," eds. Jaak Panksepp et al., 2012. http://fcmconference.org/img/CambridgeDeclarationOnConsciousness.pdf.

Chapter Six: Armed—and Roboticized

151 **PETA is not exactly:** "Eight Legs: Good; Two Restaurants: Bad," *The PETA Files*, 2010. http://www.peta.org/b/thepetafiles/archive/2010/04/23/Eight-Legs-Good-Two-Restaurants-Bad.aspx.

151 **PETA has also made:** Darren Rovell, "Octopus-throwing Tradition Still Has Legs," *ESPN Sports Business*, 2002. http://espn.go.com/sportsbusiness/s/2002/0522/1385601.html.

152 **into a vertebrate-like appendage:** German Sumbre, Graziano Florito, Tamar Flash, and Binyamin Hochner, "Octopuses Use a Human-like Strategy to Control Precise Point-to-Point Arm Movements," *Current Biology*, 2006. http://www.cell.com/current-biology/abstract/S0960-9822(06)01274-7.

155 **This regenerative ability was first:** Mathilde Lange, "On the Regeneration and Finer Structure of the Arms of the Cephalopods," *Journal of Experimental Zoology*, 1920. http://onlinelibrary.wiley.com/doi/10.1002/jez.1400310102/abstract.

155 **but Heather Bennett:** Steven Spearie, "Does the Octopus Hold the Secret to Human Limb Regeneration?" *The State Journal-Register*, 2010. http://www.behealthyspringfield.com/sections/local-news/does-the-octopus-hold-the-secret-to-human-limb-regeneration.

157 **As Mather and her:** Jennifer Mather, Roland Anderson, James Wood, *Octopus: The Ocean's Intelligent Invertebrate* (Timber Press, 2010), 82.

161 **The self-proclaimed "OCTOPUS Project":** OCTOPUS results. http://www.octopus-project.eu/results.html.

166 **the walking problem:** Marcello Calisti, Michele Giorelli, Guy Levy, Barbara Mazzolai, Binyamin Hochner, Cecilia Laschi, and Paolo Darlo, "An Octopus-Bioinspired Solution to Movement and Manipulation for Soft Robots," *Bioinspiration & Biommetics*, 2011. http://www.octopus.huji.ac.il/site/articles/Calisti-2011.pdf.

Chapter Seven: Hunting

174 **Octopuses eat about a third:** Jennifer Mather, Roland Anderson, and James Wood, *Octopus: The Ocean's Intelligent Invertebrate* (Timber Press, 2010), 57.

174 **As Aristotle noted in:** Aristotle, *History of Animals*, Kindle edition, 273.

175 **As Aristotle observed:** Ibid.

177 **Thanks in part to:** Mather et al., 124.

179 **some 69 percent of genes:** Atsushi Ogura, Kazuho Ikeo, and Takashi Gojobori, "Comparative Analysis of Gene Expression for Convergent Evolution of Camera Eye Between Octopus and Human," *Genome Research*, 2004. http://genome.cshlp.org/content/14/8/1555.long.

180 **Albert Titus, an engineer:** Albert Titus and Timothy Drabik, "Optical Output Silicon Retina Chip," *SPIE Proceedings, Optical Engineering*, 2003.

180 **our cumbersome curved-lens:** J. Shirk, M. Sandrock, D. Scribner, E. Fleet, R. Stroman, E. Baer, and A. Hiltner, "Biomimetic Gradient Index (GRIN) Lenses," *National Research Lab Review "Featured Research,"* 2006. http://www.nrl.navy.mil/content_images/06FA1.pdf.

183 **Hong Yong Yan, a physiologist:** Marian Hu, Hong Yong Yan, Wen-Sung Chung, Jen-Chieh Shiao, and Pung-Pung Hwang, "Acoustically Evoked Potentials in Two Cephalopods Inferred Using the Auditory Brainstem Response (ABS) Approach," *Comparative Biology and Physiology, Part A*, 2009. http://ntur.lib.ntu.edu.tw/bitstream/246246/162905/1/25.pdf.

183 **Other research has suggested:** F. Hart, R. Toll, N. Berner, and N. Bennett, "The Low-Frequency Dielectric Properties of Octopus Arm Muscle Measured *In Vivo*," *Physics in Medicine and Biology*, 1996. http://iopscience.iop.org/0031-9155/41/10/013.

184 **octopuses, like dolphins and:** Michel André, Marta Solé, Marc Lenoir, Mercè Durfort, Carme Quero, Alex Mas, Antoni Lombarte, Mike van der Schaar, Manel López-Bejar, Maria Morell, Serge Zaugg, and Ludwig Houégnigan, "Low-Frequency Sounds Induce Acoustic Trauma in Cephalopods," *Frontiers in Ecology and the Environment*," 2011. http://list.esa.org/pdfs/Andre.pdf.

184 **be able to make noise:** Angel Guerra, Xavier Martinell, Ángel González, Michael Vecchione, Joaquin Gracia, and Jordi Martinell, "A New Noise Detected in the Ocean," *Journal of the Marine Biological Association of the*

United Kingdom, 2007. http://journals.cambridge.org/action/displayAbstract?fromPage=online&aid=1367488.

186 **Hochner, Kuba, and their colleagues:** Tmar Gutnick, Ruth Byrne, Binyamin Hochner, and Michael Kuba, "*Octopus vulgaris* Uses Visual Information to Determine the Location of its Arms," *Current Biology*, 2011. http://www.cell.com/current-biology/abstract/S0960-9822(11)00108-4.

189 **One expedition to the Antarctic:** E. Undheim, D. Georgieva, H. Thoen, J. Norman, J. Mork, C. Betzel, and Brian Fry, "Venom on Ice: First insights into Antarctic octopus venoms," *Toxicon*, 2010 .http://www.ncbi.nlm.nih.gov/pubmed/20600223.

190 **This obscuring "membranous":** Henry Lee, "The Octopus and His Prey," originally from *Land and Water*, reprinted in *The New York Times*, 1873. http://query.nytimes.com/mem/archive-free/pdf?res=9D05E1D8103CE731A25755C2A96E9C946290D7CF.

Chapter Eight: Sex and Death

194 **Aristotle has a few:** Aristotle, *History of Animals*, Kindle edition, 90.

196 **not a gentle organ:** Aristotle, 88.

197 **Christine Huffard, a biologist:** Christine Huffard, "Ethogram of *Abdopus aculaetus* (d'Orbigny, 1834) (Cephalopoda: Octopodidae): Can Behavioral Characters Inform Octopodid Taxonomy and Systematics?" *Journal of Molluscan Studies*, 2007. http://mollus.oxfordjournals.org/content/73/2/185.full.

202 **As Aristotle so poetically:** Aristotle, 125.

202 **From now on:** Aristotle, 138.

203 **Roy Caldwell, of the University:** Roy Caldwell, "Death in a Pretty Package: The Blue-ringed Octopus," *The Cephalopod Page*, reprinted from *Freshwater and Marine Aquarium*, 2000. http://www.thecephalopodpage.org/bluering1.php.

204 **it is a genetically programmed:** Jennifer Mather, Roland Anderson, James Wood, *Octopus: The Ocean's Intelligent Invertebrate* (Timber Press, 2010), 147.

204 **Even initially dismissive:** Aristotle, 273.

Epilogue

213 **a recent analysis says:** Jessica André, Malcolm Haddon, Gretta Pecl, "Modeling Climate-Change-Induced Nonlinear Thresholds in Cephalopod Population Dynamics," *Global Change Biology*, 2010. http://se-server.ethz.ch/Staff/af/Fi159/A/An287.pdf.

INDEX

Page numbers in *italics* refer to illustrations.